STRING COLLIZION:

Just a Collection of Strings

By Chris Leermakers

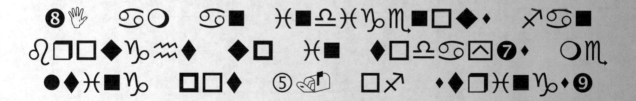

"I am an indigenous fan brought up in
today's melting pot …. of strings"

"When Old Ideas Collide, New Ideas Emerge"
Christopher John Leermakers.

Balboa Press books may be ordered through booksellers or by contacting:

Balboa Press
A Division of Hay House
1663 Liberty Drive
Bloomington, IN 47403
www.balboapress.com
1 (877) 407-4847

ISBN: 978-1-5043-1189-2 (sc)
ISBN: 978-1-5043-1190-8 (e)

Print information available on the last page.

Balboa Press rev. date: 01/19/2018

BALBOA.
PRESS
A DIVISION OF HAY HOUSE

FOREWORD

Welcome
Hi there and welcome.
You've entered my hypothetical world.
Such delicacies and delights!
You'll never leave my world.
When you come to visit
I'll give you more to drink.
But if you seek some profit,
I'll send you to the brink.
I play the fool too often,
Dance and sing regret.
But when you are near and with me
How soon I forget.
Here it comes again
I hate the question Why
Just wait until we're dancing
Your eyes would wish they'd cry
Pleasures of the great unknown
I do not want to find
Because here in reality
There is no other kind.
Sleep and wake tomorrow,
See the brand new day
To hell with all the sorrow,
Help us find a way.

Beauty is among us
We see it all the time.
Let's see if it ferments
And drink it like wine.
By now it is too late
The grape is getting wrinkly
And how you hesitate
To see your future quickly.
Oh well, what can we do
We are intelligent animals.
I am just like you
Death is a natural progression.
I don't have to like it,
Life is a frustration
Filled with little profit.
When you see the Sacred One
Your eyes begin to glitter
The heart pushes the buttons
And the mind is filled with litter.
Don't ever say never
Try not to be discrete
Because in my hypothetical world
We always love to meet.

CHAPTERS:

ACT ONE

ABSINTHIUM

TRY

Away run to try can you
Play and come in to try can you
Go me let to try can you
All at you help it won't but
Fall you when there be will I
This from away far, away far you take to try to…
Hand the by you take me let
See you make to try like and
Know to is really there what
Free for stay can things what and
Me with along come you will
Long to stay not will I for
Dance and drink and eat can we

Song a sing even I'll and
Road a is there life to
Comes it as take can you
Path the from stray may you
Seek you what is truth if but
Despair into cry might you
Fair never was life but
Place this to me brings what it's
Waste to mine and domain my
Help to try even can you
Forget, purge and heal to try
Use no is Really there but
Abuse mine's of trait great a
Lose to left nothing have you
Try to but….

A NEW STATE

I'm back into, entering the maze
Or something new to find
Yes, just passing through?
Just passing through
The next phase
We wonder
Wonder when our time has come
Some say it will take forever
Could be true
But we all have seen a thing or 3
Sometimes black stars shine
Am I the only one? - I doubt that
It could be put on my epitaph.
Survival always strong, strong at the start
Lifelong, stronger in the heart
Whatever it is
Whenever it rises – the new state
Will always stay, always be
Ready I say
No matter what it is
The new state will always get to play
Ticking
Time ticking sometimes feels
Like a rock
Always looking for new state to dock
Cities by the sea?
Sounds good to me
There is no bitter end
Just a better end
The days are still hot
Some nights are still cold
For we are all just here
Waiting for the new state
Watching it all unfold

I just got out of that last maze
It was just something
That could never work
Something just like a hazy
Memory, something we already solved
Who we really are
From what I've been told
Who we really are?
You haven't seen nothing yet
No we are not just one history scar
In the new state
We should smoke a cigar?
Cause if our time has really come
Are you ready for it all?
Or will you just fail
Fall
Will you wail
When this world falls apart?
Yes when your world falls apart?
Not everyone will see it all through
Maybe not even me or you
But one thing I know is sure
There will be enough last one's standing
To make this all work pure
Last ones standing
Never an easy job, but just remember
The last ones standing will be there
When the corrupt fall apart
When the false churches fall apart
When the U.S Of A begs to be saved
Yes, when the old New World Order
Is taken down
When the New Age takes its crown
Takes over from all the older fools

Who thought they were the clever at the start
The elite
Won't want to let go from their savage bite
Not a pretty sight - a lot will cry
And maybe, a lot will die
Although in the end, true "God" (Mathematics,
Physics, Science, Youth & Intelligence)
Will win their plight
Yeah will end the whole, but bloody
Fight
After that, not much left
Those in power give the cue
For the Venus Project to start
Designing the future – looking forward
And ahead of their own time
To shine and rebuild our timeline
And it's about NOW
Yes NOW is the right time
For it is a new state
Already thought of
Already plans drawn
Drawn up many years ago
Give your thanks this time
Give your credit – please guess
Who saved your life – new chances? - yes I think so
New stances
New chances for all our children
Yes yours and mine

AEON=AGE: AGE=AEON

Never really new
From the beginning
Looking up at the stars
Yes the stars were shining
Running around like children do
But at night roaming the streets
Looking for new friends to meet
Our sun is still shining for now
Earth is on the right tilt
On its right axis
But it really doesn't take much
To distract us
Some say
An extinction level
Event
Is coming
Some pray
For some of this special
And deadly rain
To wash away our world's pain
And some claim
They know the date, but
They are just foolish or insane
No one knows our true fate
Every aeon
Every age
Just natural
Earth changes?
Some are man made
Changes in space right above
But out

Right out of our control
Scaring some right
Through their soul
Some call it The New Age
I just call it a progressive stage
Maybe it is the end
The end of this age
But I think we're ready
For this new stage
A new stage of intelligence we will follow
The fake sun God is almost finished
As it even said 'I'll be with you
Until the end of The Age'
Remember age equals Aeon and Aeon equals Age
All evolving, all receiving
Looking forward
Not backward
We're the ones who survive
Each Aeon or Age the real universal God
Has survived
Not afraid of any new stage
This is the nature of things
It's what it can bring
Every so often
Something erupts in the sky
Erupts in outer space
But on Earth we can get
A dangerous light show
Yes a dazzling and magnetic display
But dangerous magic embrace

BLIND FAITH

Do you know what is coming?
Mirror cracks appearing
Isn't it so revealing
Revealing
Can you see it?
I don't think so
Can you feel it?
We don't know
But you're still here
Not letting all go
Never letting go until
You fall into your pit.
The weak, the meek
Won't fold
Yeah won't choke
You won't hear any cries
I've got the ace of spades
And the king of hearts
Always had the upper hand
While you were born
With your head in the sand
We won't choke
Choke on your force fed lies
Please see yourself
Through your own blood shot eyes.
Another silly test?
I'm not your enemy
Or the devil in disguise.
So funny really
Or do you even recognise
Who it is, what it is
Confusing you?
Don't you want
My bull-shit proof vest?

Speaking truths is not easy for some
And sometimes it's not meant to be
And doesn't come free
But it's just so easy for me
It's just what I see
I'm not a prophet
I'm just me.
Yes, I'm actually
Free.
Betrayal and self denial are easy
Have you lost
Yourself again,
Led yourself astray?
Not me
Keep your criticisms to yourself.
No-one is listening,
No-one is waiting.
Everyone has stopped believing
In your false God story today.
It's your blind faith,
Keep your own blind faith.
Blind faith
Stop stealing from the living
..and the dead
We're not interested
In what you said
Just keep following
Yeah, keep swallowing
Your own blind faith.
Hook, line, fool and sinker
Don't you want the real God,
The real Creator on your side?
It's just your own stupid pride.
Silly monkey pride.

Trix are for kids
The young and the free will dictate your future,
Older generations had their time.
Yeah, had their turn
And we nearly all burned.
Like rabbits running
And trapped in a hole,
There is no such thing
As control.
Just a deluded memory
Your delusion,
Yeah, just your own mind control
Your own confusion
Just misery
It's just your big fucking ego
Consuming you
Just let it go.
Are those cracks
All mirror cracks
Re-appearing?
Can you see it?
I think so.
Can you smell it?
We don't know.
Cause I won't choke
On any force fed lies.
It's your own blind faith
The root of all evil
And gladly not mine.
I'm no prophet
I can just see.
Yes, I'm just me
The free
Just me.
Yeah always was and still
Forever free from all
People like you,

OUTSKIRTS

I might have cried in another life
Yes that would be right
Seen it all before
Not a pretty sight
Yeah, not a pretty sight
On the outskirts of life
Many are quick to judge
Many are just fools and jokers
Many don't have the right tool
Yes, the right tools in hand
But you need them
In this world wild land
The outskirts of life
Oh, what a story to tell
Outskirts of life?
Yes we already fell
Better pull our heads in
Don't shit in our own yard
Cause' no one will look
To trust you - even from a world apart
Stopping digging your own grave sites
Start planting some flowers
Even if they go really sour
Positive energy is all you need
Yes positive energy is what you must feed
Or the
Outskirts may just come back
The outskirts of life -

Just welcome them back?
The outskirts are my home
Sometimes a unique state of mind
It can be something also new, I find
Now that we are in a later
Stage of time
The outskirts might just free our minds
I know we've cried in another life
Yes that's right
Seen that darkness before,
But now it's a pretty sight
A very normal site
On the outskirts of life
Some viewpoints make me want to give
A certain shade of light
It allows you to see my illuminated life
But don't change on my behalf
You do what you will
Do what you want
You can do anything
In order to make yourself right
To give you a taste of righteous life
Are the outskirts my home?
Yeah it's home to me
Because I'm not just happy
I'm free

THE REAL ILL

Our men and women put their hands up
Hands up all over the world
Hands up high to defend our lands
Defend our freedom .and our lives
Yeah also our cries
They are now returning home
Some hurt and broken
Some dead, and some hoping
But most have now just awoken
Yeah now they have spoken:
Many lives, yeah many lies
Lost and destroyed
A lot of our 'real' targets were just plain missing
Just some ones' evil imagination
Yeah so many lives, so many lies
But not near enough real cries
Some of our orders just seemed fake
Yeah just like they came out of a cake
Yeah like plastic, and some just outright spastic
We would, yeah we still will
Kill for our freedoms
But only for the right reasons
Not more and no more for those
Yeah those manipulating, deceiving little flies
Yeah no more for all these desecrating and deadly lies
When they return
Some things are never the same
Nothing is really returned
Back to them
Sometimes just blame
And none are looking for

Any fame
Yeah none, no more are looking to play
Their evil master's game
Some families get nothing
Or nothing the same, nothing to gain
Just a blood stained flag
Yeah just a burnt flag
Yeah some things are never again
The same
All were just stars awaiting
Awaiting to make their country, their family proud
Yeah all were stars just waiting
To wave their flag celebrating
Yeah sorry, but now all they see is the truth
Yeah now they're all, all just stars
And scars
Yeah scars on a bleeding flag
Scars on a burning flag
All took oaths to defend our countries
From all enemies: foreign and domestic
After the innocent blood seen
Yeah after the innocent blood spills
Yeah now - all knowing their own leaders are the ill
Yeah own leaders: the real enemy, own leaders are the real kill
Own leaders: the real true 'missing' targets
Yeah this time it is no one's imagination
But just in those living in damnation
Yeah our own leaders, We won't forget the real ill
Yeah we will never forget the huge hole
Yeah the huge hole you tried so hard to fill...

BAD BIRDS (BIRD IS NOT THE REAL WORD)

It's getting close - any time now
By the way
Don't ask me how
Sometimes moving your shadow
Let us soon forget
On how we met
Once in a million? – no better yet

Where did you go?
What did I do?
Didn't I tell you the truth?
Who's to know – the better clue?
You?

What the hell happened
In the end, I broke that spell
It lasted – long
Long is the word, it was a bad bird
Yes, just ask that colourful Humming Bird
Cause' this bird knows that
Bird is not the real word

I should have stopped that bad bird
When I had the chance
Yeah I told you
I broke its spell
Why am I in the wrong? – Because
I've spoken the wrong words?
Cause I spent some shells?

I'm Ok now - always clear
Near here - or far
Even at my favourite local bar

I still should have shot that bad bird
Instead I clipped its wings
Dropping to the ground
Into his own shit - but just gave him a truth sting

Should I shoot that blackbird?
Will I somehow regret?
Maybe he now
Wants some more
Trying to settle his dark score?

He's really close now
But I aint moving
Anywhere, anymore
Neither is my shadow
Where did he go – to hell and back?
Who really knows
I just don't care if he hangs still
Even with the good crows

Well now I've healed that bad bird
It was a good hit
Dropped to the ground
This time not missing
His bottomless pit.

JUST ONE BEER

I've had it up to here
It's amounted to so much
Strange noises in the distance
I'm starting to get fed up
Hey tell me what's the answer
What's it all about
Frustration building in the air
In this dead end shout

Looks like it's that time again
Can't even have a beer
Can't stand it any longer
I've gotta get out of here

Looking for some action
What do I want tonight
I roll up in a dusty bar
And drink 'til I'm alright
Then some dude comes on over
He's looking for some action
Yeah looking for a fight
I can't even get happy
So I'll just put him asleep
Yeah nap time is just early

For him tonight

Looks like it's that time again
Can't even have a beer
Can't stand it any longer
I've got to get out of here

So I get my head together
It's time to have some fun
I see some dude selling DMT
I promise mum, just one
Soon enough I'm spinning out
What brought me to this place
See strangers all around me
And such an evil face

Looks like it's that time again
Can't even have just one beer
Can't stand it any longer
Just saw the police outside
Yep, looking stronger
How am I gunna get out of here....

BALANCING

I start to feel it too
Strange feelings all over me
Now it's over me
Now it's over you
Come on, let me see
The dark soon glides by
Our park passed the sky
If it goes, you do not cry
Over, the passing by
Life revolves
Around the clock
And then evolves
Breaking the lock
To them divine
To me beauty
Stand and guard
Painstaking duty
And soon the air
It drifts into me
You are there
And I am me

Floating high
We are free
In the sky
Catch me please
Don't you freeze
On me now
More than ever
Watch me how
Balancing
In the Spring
I see the air
Over there
And watch it sing
With everything
The blooming blue
Is after you
It's running fast
It's running past
The likes of me
I'm coming last...

HAVE ONE ON ME

I'm still thinking
Thinking about last night
Yes still dreaming
Dreaming a tomorrow's dream
Have you forgotten
Forgotten tomorrow
Forgotten me
Everything already?

Yeah we're not tired, just a bit wired
No not tired, no way, not dead
Yet,
So have one on me
Just have another one on me
And I'll have three
Angel I said
Just have one more on me

Morning is nearly here
I should get some sleep
Before the sun hits our room
But you're still too beautiful
Too beautiful for me
So we better lock the door soon
Yes, welcome to the boom.

Sometimes I can see past your heart
And feel your love inside

Don't get embarrassed
Just show your pride
And have one on me
Just have one Absinthe on me
Yeah just have a Jager on me
Yep, now we're back to the start
Yeah we're still awake
Making love – half asleep
Laughing and screaming
Why not another one on me
She had another one on me
Am I still dreaming?
Dreaming about last night
Yeah still thinking
Thinking about yesterday
And
I know you haven't forgotten about me
Haven't forgotten how to see

You're still so beautiful
So beautiful - an angel to me
Still seeing beyond your heart
See her love inside, now the Absinthe
Is turning it all inside out
Be proud my angel and have one more
On me
But only if you can see!

BAD BIRDS : ACT 2
(Is Bird The True Word?)

It's getting close
Any time now
By the way
Don't watch me now
Let's soon forget
Of how we met
Once in a zillion
Better yet
Where did it go
What did I do
For who's to know
The better clue
Hey, then again
What the hell
In the end
Broke the spell
It lasted long
With a word

And a Song
I killed the sad bird
Why am I
In the bad
I nearly cry
But doesn't make me sad
Hell I don't know
Where I'm to go
Near or far
My favourite bar
Still
I'll sink a few
For what's it due
To, I don't care
The burning of
My long dark stare...

BLACK SHADOW

I'm in no race today
To take your place
In no hurry
To follow your daytime charade
That dark shadow
Yeah no shadow, nothing in my face
This day
And this I contain…just let me explain
No one needs your dark faces
Looking down
Anywhere around
Everybody's already aeen
Yeah everyone knows
Owns
Personal dark places
Their own dark faces
Even on their ground
I have found
Just follow your own black shadow
Yeah your own black shadow
Needs, yeah feeds
Owns only you
Only you today
This I am sorry to say
The black shadow comes crawling

Crawling back to your door
Black shadow, back to shock
And back to lock, lock you again
Yeah to nail your head to the floor
Your black shadow comes running
Always running, always ruining
Yeah ruining your core
Returning your call
Black shadow comes running
Always back for more
Even if he is the only one to fall
Why did you go back for more
How could you
Yeah how did you open his door
Try to look forward my friend
Not just straight ahead
Even if it sounds like a lie
A black shadow will always try
Try to confuse
Try to consume
If you listen to it all and follow his call
If you're listening to the end
You just might end up dead
And your black shadow will just
Replace your head

KEEP TRYING

You can try to run away
You can try to come and play
You can try to let me go
You can steal me from the show
But it won't help you at all
I will Be there when you fall
To try to take you far away, far away from
This...

Let me take you by the hand
And like try to make you see
What there really is to know
And what things can stay for free
Will you come along with me
For I will not stay to long
We can eat and drink and dance
And I'll even sing a song

To life there is a road
You can take it as it comes
You may stray from the path
But if truth is what you seek
You might cry into despair
But life was never fair
It's what brings me to this place
My domain and mine to waste
You can even try to help
Try to heal and purge, forget
But there really is no use
A great trait of mine's abuse
You have nothing left to lose
But to TRY...

9*1=9
9*2=18(1+8)=9
9*3=27
9*4=36
9*5=45
9*6=54
9*7=63
9*8=72
9*9=81
9*10=90
9*11=99
9*12=108
9*13=117

MIRROR1} The True One

Input:
0…..10
1…. 24
1……4

Output:
3
1…..(3)
1___(3)

REMEMBER THAT ZERO (0) IS NOT A NUMBER

Table 1

1	2	3	4	5	6	7	8	9
A	B	C	D	E	F	G	H	I
J	K	L	M	N	O	P	Q	R
S	T	U	V	W	X	Y	Z	

Table 2

1	2	3	4	5	6	7	8	9	10	11
A	B	C	D	E	F	G	H	I	J	K
1	2	3	4	5	6	7	8	9	10	11
12	13	14	15	16	17	18	19	20	21	22
L	M	N	O	P	Q	R	S	T	U	V
W	X	Y	Z							
23	24	25	26							

0NE=THREE & THREE=ONE (1=3/3=1)

1=3/3=1
2=4/4=2
3=5/5=3
4=6/6=4
5=7/7=5
6=8/8=6
7=9/9=7
8=10/10=8
9=11/11=9
10=12/12=10
11=13/13=11

```
        11:11
[8]1=3/3=1        [8]
    2=4/4=2
    3=5/5=3
    4=6/6=4
    5=7/7=5
    6=8/8=6
    7=9/9=7
    8=10/10=8                    81[9]/18[9]/81[9]/18[9]
    9=11/11=9        4*9=36(9)   9x9x9x9 = 6561
    10=12/12=10                  11:11 = Love won
[8]11=13/13=11  [8]
    11:11
```

Multiverse Theory=
4332945915 2856971

MULTIVERSE THEORY (M THEORY) IS PROVEN BEFORE THE 5TH DIMENSION (3=5/5=3)

(1)   GOD/LOVING, INFINITE & EVOLVING MULTIVERSE
(2)   CHRIST/TRUTH MESSENGER/TRUE SPEAKER
(3)   HOLY SPIRIT/SOUL KNOWLEDGE
(4)   MATHEMATICS
(5)   PHYSICS
(6)   SCIENCE
(7)   ASTRONOMY
(8)   ASTROLOGY
(9)   ART
(10)   MUSIC
(11)   INDIGENOUS PEOPLE/DREAMTIMES
(12)   D.N.A

1+2= (3) (6) (9) > 12 =

INFINITE & EVOLVING DNA/UNIVERSE &MULTIVERSE

CHRISTOPHER JOHN LEERMAKERS
(67) (20) (44) = 131(5)/313(7)/333(9)
D.O.B: 13/10/1974=
1/3/1/0/1/9/7/4 = 26 (2+6)
9= MULTIVERSE
10=MUSIC

TRUE NUMBERS REGARDING DRUGS/MEDICATIONS/NATURAL HERBS AND NATURAL COMPOUNDS

(1)   WORMWOOD = (56945664)=45 (4+5)= 9

(2)   CANNABIS = (31551291)=27 (2+7)= 9

(3)   KHAT = (2182) = 13/31/33 (3X3)= 9

(4)   D.M.T = (442) = 10/1 = DIMETHYLTRYPTAMINE = 93 (9+3)=12

(5)   THURJINE= 28391955=42/METHERGINE=4528597955=(59) (42+59)=101/303/11/33 (3X3)=
        9(5+9)=14/(3/6) 3+6= 9

(6)   ABSINTHE(HERBAL SPIRIT)=33 (3X3)= 9

(7)   JAGERMEISTER(HERBAL LIQUEUR)=58 (5+8)=13/31/33
        1+3+3+1+3+3+3+1=18 (1+8)= 9

(1)   KHAT=2812 = 13/31/-33 (3X3)= 9
        "GOD IS A NUMBER THAT YOU CANNOT COUNT TO" Marilyn Manson (Bryan Hugh Warner)
        "I DON'T LIKE THE DRUGS BUT THE DRUGS LIKE ME" Marilyn Manson (Bryan Hugh Warner)

REAL VORTEX MATHS & TRUE MIRROR MATHS & The Mystery Of The Number Sequence: 3 6 9

(ZERO (0) IS NOT A NUMBER)

$3 \times 1 = 3 \, (1x \quad) = 1$

$3 \times 2 = 6 \, (6+3)=9 \, (6X3)=18 \, (6+5)=11 \, (11=9) \, (1+8)=9$

$3 \times 3 = 9$

$3 \times 4 = 12 \, (1+2)=3 \, (12-3)=9$

$3 \times 5 = 15 \, (1+5)=6 (15=37) \, 3+7=1 \, (3) \, \& \, (5) \, \& \, (333) = 3+3+3= 9$

$3 \times 6 = 18 (1+8)= 9$

$3 \times 7 = 21 \, (2+1)=3 \, (3+7)=1 \, (33) \, (3X3)=9$

$3 \times 8 = 24 \, (2+4)=6 \, (24+6)=3 \, (6+3) = 9$

$3 \times 9 = 27 \, (2+7)= 9$

$3 \times 10 = 30 \, (3+0)= 3$

$3 \times 11 = 33 \, (3+3)=6$

$3 \times 12 = 36 \, (3+6)=9$

$3 \times 13 = 39 \, (3+9)=12 \, (3x9)=27(2+7)=9 \, (1+2)=3 \, (12-3)=9$

{1}1999 = 28 (2+8)=(1)<(1)   (2)   (3)   (4)   (5<

2) 2000 (6) 2004 (10)/1 2008 (14)/5 2012 (18) 2016 (9<

(3) 2001 (7) 2005 (11)/2 2009 (15)/6 2013 (19) 2017 (10<

(4) 2002 (8) 2006 (12)/3 2010 (16)/7 2014 (20) 2018 (11<

(5) 2003 (9) 2007 (13)/4 2011 (17)/8 2015 (21) 2019 (12<

9/11/2001 WAS AN INSIDE JOB USING BOTH 3D & 4D/5D TECHNOLOGY.

1. LOVE
2. TRUE/TRUTH
3. SOUL KNOWLEDGE

The First 12 Fibranachi Numbers and relationship to 3 6 9

1	1	2	3	5	8	13	21	34	55	89	144
1	2	3	4	5	6	7	8	9	10	11	12
1	1	2	3	5	8	4	3	7	1	8	9
1	1	2	3	5	8	4	3	7	1	8	9
2	3	6	9	1	2	4	6	1	1	2	4
3	7	1	8	9	8	8	6	4	5	6	6
3	7	0	7	7	4	1	5	6	1	7	8
8	8	7	6	4	1	5	6	2	8	1	9
1	1	2	5	8	4	3	7	1	8	9	
8	8	7	4	4	5	2	8	1	9		
9											
9	9	9	9	9	9	9	9	9	9	9	9

Reduce all 12 (label beside row 2)

Reduce all to 9 (label beside row 10)

$12^2 = 144$

First in series with root of 9

Find the next 9 in the series

$4657 = 27 \ (2+7) = 9$

Infinite pattern emerges 999

Vortex based mathematics reveals a high dimensional symmetry.

THE EVIL ILLUMINATI STILL MIGHT BE PLANNING ANOTHER 9/11/ FALSE FLAG TYPE OF EVENT BUT ON A MUCH
LARGER SCALE TO TRY AND TRICK THE WORLDS' PUBLIC ONE LAST TIME. BE VERY AWARE.

Cheers,

Chris.

ACT TWO

THE GREEN DRAGON:

Standing out and looking
I hold you close tonight
Wondering and thinking
Reality's out of sight
If and when I see you
I won't know what to say
Seeing you is better
I'll remember you that way
Saying words to please me
You tear my heart apart
Feelings, now they're easy
Especially from the heart
Life is filled with choices
We see these every day
I hear the same old voices
But don't know what to say
Take them all together
They wind up best this way
Take them all forever
Immortality's here to stay
Waiting for the winter
To one day go away

You appreciate the summer
Funny, isn't it I say
Let's drink the wine together
It's ours and ours to drink
And soon the cool sweet water
The best drink, don't you think
Add to this a lemon a little Absinthe too
It's pale yet so refreshing
Ooh, here comes a green dragon
I think he'll find it quenching
Drink it all night long
We'll drink them one and all
Then maybe sing a song
Retire, then rest, then crawl
Maybe grow some fruit
And tend to the placid sheep
Trip over the green Wormwood root
So we can one day sleep
People, places, memories
Experiences quite unique
Remember them as always
For one day we will meet…

YOUR VERSION:

Tell me of your world
And I'll tell you of mine
Increasingly unpopular
The black light divine
Where the sun hits down hard
From the Earth to the Moon
Tell me of your world
And I'll tell you of mine

Look up
Tell me what you see
Are there eyes looking down
In the forest of night
Can you see the white lagoon
Within the limits of your sight
So there the creatures sit
And wait for your command
Wondering for a moment
Impatient thus they hunger
Confusion rules the land
Next emotion anger
Bitterness surrounds us when sweetness is a lie
I can see them laughing
In the corner of my eye

In time they will discover
The wrongs that have been done
Let them stay and take cover
I'm heading for the sun
The place where I reside
Fresh inexperienced fruit
Make me want to hide
Here is only freedom
And beauty all around
Nothing to distract you
Or move you from your ground
We see ourselves as children
Stuck in an adult frame
Every day is different
And nothing stays the same
The path to life is simple
You've heard it all before
I want to hear your version
So come and tell me more
It may not be short and sweet
Or in another time
Tell me of your world
And I'll tell you of mine.

SIR X:

My Existence is truly enigmatic
I can't work it all out
Why i'm here
What's it all about
Take dreams for example
Our Minds Twisted
Our souls, who's to know
And though we are retarded
I wonder if they'll show
So many things unknown
Influences vast and few
The seeds have been sewn
Intoxicate with you
But No, It's in our blood
In one day
Explain the crude
DNA

Oh, how do you know
I mean, are you sure
Which way to go
Oh mr., how do you know
The bureaucrats
They're little mice
Like little weeds
Fighting lice

Oh well, they come and go
Making a name
On with the show

And in the end
They still Pretend
In order to score
A little more
Save me some
Sir X

I think i'll come
Madam Sex
What's it all about'
Can you work it out
Questions here and there
Solutions everywhere
So we want to look
I think i'll read a book

I do not want to look
Instead i'll read a book
What's it all about
Can you work it out
Questions here and there
Solutions everywhere...

STRING COLLIZION

Now you have your marching orders
Can you still run? - no?
Can you still walk? - no?
Do you see the disorder?
That you have left behind
And now do you, how do you even sleep
No, I'm not that kind!
No, I'm not one of your sheep

Tell me how does it feel
When the puppet strings are strong
Yes, when the puppet strings are long?
Now you've felt their strength
You even might have
A vision?
That you're on your own collision
Your own string collision

Have you still got your ego?
Didn't it start aeons ago?
Did you have any power?
Maybe it went all sour
You think
Still think you have any control?
You just woke up
Now you know
What

Everybody already knows
You are the mole
Yes, the inside mole
That came out of a dark hole

It's your own break down, every day
Yes, we have learned how to see
Learned how to be free
To stay away, yes
It's your own big melee,
Your own car crash today

On a collision course
Within a big pool
In a deep pool of fools
Yeah, it's not my scene
Aint never going to be
That's all that I see
You're the one on the real collizion course
A collizion course with me
A real collizion course with the free

Your marching orders kept you away
I know you can still walk
Know you can still talk
Know that you can still sleep
But you're still a wolf

Pretending to be a sheep

Thinking you can lead
The weak and naive astray
But we are here to remind you
That time is your real master
And that you're real the muppet
Yes, the real puppet?
Having a very bad day

So, still you're thinking
Thinking
That you live in
That you can see
Only three
Dimensions, yeah just three
Time is on my side
But you won't be able to catch me
Won't even be around to see it

As I am me,
As I am free
In my time line
And I have just been fine
Will be, until
The end of my line
Maybe until the end of your time

It might just be a theory
But maybe a theory that evolves,
Involves
Just a collection of strings
So don't you cry in front
Don't even sigh in front
Of me
I don't care if it stings
Cause' you still pretend
You can work it all out
What this is all about

No solutions, no questions
You seek? you have?
No science, no evidence
You talk? You want?
Well I think,
I will
Read a real book and have a good walk
And see real eXistenZ,
No fake - existence for me
Why don't you do the same?
Then you might have a slight
Chance in this distinct
Yeah this different
This unique time game

TRUE EXISTENZ

The balls of glowing gases
Never look the same way
They're twinkling in the sky
At a zillion miles away
Massive immovable objects
Yet moving all the time
Always getting larger
The question in the back of my mind
Imagine the great distances
That separate each lonely star
Imagine it, how can you
No matter how advanced you are
Our existence is truly enigmatic
Yes, true existence
True Existenz we all are

At night
I see the stars moving
I wonder at the sight
Or are we just spinning
In the infinite sky tonight
Hydrogen and elements found
In our domain to this day
They act as fuels for such light and sound
Life within the Milky Way
Stars stand for things far and wide
Any places, any thing
Constellations that just can't hide

Or can they, an enigmatic thing
Yet, this we do not know
Don't think we could handle it
Otherwise we'd try and go
Already, thus departed
Turn and face a different star
You don't know just how lucky you are
Are you really interested
Or has the sight
Been taken in for granted

Nothing is impossible
The difficulties and risks may increase
These can be adaptable
Otherwise our dreams will cease
Minds travel at an extraordinary pace
The hopes and dreams, they don't resist
For there is no time and space
And no boundaries that exist
To think we are alone
We'd have to be naïve
For in the Twilight Zone
They mislead to deceive
Our galaxy is but one of many
There are so many far away
Will we ever reach one?
Well maybe one day…

OUTER SPACE

Into each lonely star
No matter how desolate
And far apart they are
For we will one day know
The reaches of deep outer space
In time man can get to know
A different, intelligent and challenging
Living race
Man might think he is superior now
Or that there is no other kind
There is no other kind
No matter of the thoughts
That travel through our mind
Some things are left unknown
It works out best this way
For we don't want to wake up
In a different kind of day
I can't take for granted
It makes me want to shout
That each lonely star

Will someday scream right out
The light it takes to reach us
Like a great deal of spilt wine
Vast, intense, indefinite, quick
It's traveling
Through time and space
I think I'm going to be sick
I wonder if we ever knew
Would we ever know
I wonder if they do
Where can they go
Without being seen
Mars, Pluto, Jupiter
Big deal, just a different scene
Come on
Do you want to make me proud
Finally
Of being part of this race
Go out there and see
What is in outer space.

ONE DAY:

Funny little bureaucrats
Little pencil pushers
Little funny aristocrats
What's it all worth
A dollar today
Tomorrow thank you
What did you say
Oh sorry, bless you
Twisting words of logic
They think they sound alright
But one day, the nightmares cry
All through the night
They think that they are lonely
They think that they're alright
Whatever happened to homely
And I'm out of sight
Public Servants
Servants for the public
Public nuisance
Funny republic
But whoa, hang on
Just like everything else
Good and bad
We wind up sad
Then perhaps glad
Because
If there is no glad
Then everyone is mad.

DISGUISE

Watch out some times my friends
Look out
Watch out for friends in disguise
It's a real sanity killer
And you might not
Even realize.

A new day is calling
Yes a new day is coming
Now your eyes are open
Yes, now your mind is open
No more falling
Just better gathering
No more falling for the lies.

Always watch out
For that
Sweet, sweet wine
Strange and beautiful people standing in line
Just watch – you will see them in time
Yeah, no more fooling
No more fooling with your mind

Some disguises seem to fit well
Yet so revealing
When you ring their bell
These people don't concern me
Anymore,
For they are stuck, already sleeping

In their own rubbish bin
Forever more

I've been bitten
As have many, you see
But we have survived
And now ready and waiting in our good shoes
In case they surface again and try round two.
Just watch
Out sometimes
Watch out for the disguise
I know it's disturbing
It can turn you inside out
But a new day
A new way has been found
Yeah it's so profound
Yeah just shout
Our mind's eye is wide open

Defending
Defeating all real enemies
With our own two eyes
Cause when it's all over
Then realize, you will
That there is really no disguise

DEEPER DISGUISE

Have you ever heard
Of the strange yet, beautiful one
It can be a thriller
Maybe a sanity killer
You don't realise
Things I haven't dreamt
Of ever seeing at all
Strange and beautiful people
To me they sometimes seem
Yeah seem quite small
Practicing with such delights
We dance on the cutting edge
For here there are so many lights
And I will try and set a pledge
Why is this world too real
Escape is what we need
We'll even try and steal
We are a different breed

Make love you sweet one
Try to play with fire
The world today as we know it
Has turned many into a liar
Or has it turned me inside out
Or maybe

Make me want to shout
The birds making chanting calls
This fine, fine morning
Make the evening balls
A new day's calling
With Summer, Spring and Autumn too
We never have no winter
I'll sing and dance
And drink with you
And stay right in this glamour
Make the fine
The sweet, sweet wine
And let it rest
And soon digest
Like a test
We all congest
So soon take one
Take them all
But hey, watch out
The great big fall
It's the beautiful one
The friend in disguise
A real sanity killer
Or might you not or never
Realize…

HOME SWEET ROAM

I am a nomad
I walk in different places
Meet different people
And so many different faces
I wander around
Nothing better to do
Don't make a sound
Or we'll stop here too
A perpetual roam
Sets me apart from the rest
Green pastures all round
Always searching for the best
No place to go and hide
I am nothing but alive
I roam from place to place
Travel in both time and space
I am an abstract being

No challenge or regret
One who needs no freeing
Or belonging to a given set
So random and so primitive
Unattainable and so free
Like creatures with a simple mind
Irresponsible yet so delectable are we
We know not of a holiday
A profession to call work
We simply pass the day
Thinking where to lurk
On this planet, in our lives
We do have several wives
No place to call home
We simply live and roam….

HTURT SI KCAB
(TRUTH IS BACK):

It's found
Found in the back
In the back of our minds
Intelligent design
Science and sanity we will find
It's a redesign of a culture
Yes, it's back in our hands
Always backing up each demand

A firm hand is needed
The sleight of hand required
To end the distortion of our time
Time to end the hatred and lies
Some might even try to escape
Escape and fly away like flies

Watch out for the
Devils, drugs and doctors

Still trying to redesign
But they spoiled their chance last time
Just like the economic hit men
From the past
These people will never last
Cause we've always been watching
Been watching strong and fast

Welcome to a new future
Yeah welcome to the real
Structure
Real function
No more destruction my friend

All of us have a part
To kick start
To kick start our home
Kick start our world again

BREAK ME
(The Unbreakable One):

I can see you
Coming
Coming from a mile away
I can hear you thinking
Thinking every day
I can see your motive
Even in the ashtray

But it won't break me
It won't break me
Whatever you try
You can't break me, cause
You
Are my barstool
You won't break me – I'm a misfit already
Haven't you heard of that tool?

Walls can't hold me
Your doors won't stop me
Mind is clear
My mind is fine right through
But do I have
A maze for you

You won't break me

Silly one understand
In my life there is no strife
So go back to where
You're from
Go back where you really belong
We will never get along
And I think you've already heard this song

You can't break me
Can't break me down
Logic and reason
Always more powerful
But I'll throw you a bone
Just take your little clip board notes
And try and carve them in stone
In stone

Have you seen your exit yet?
Have you read your next line yet?
Haven't felt my words yet?
Cause it might just burn you
Even in a cool shade
Yeah even in the dark shade
In your dark shade

BECAUSE I'M JUST THE UNBREAKABLE ONE

Your heart of darkness
Won't confuse me
It won't use me
Cause all I need is me, you see
No one can break me
As, I am just me
Being free
And no one can break it
Break it from me

COLOURS = COLORS:

Red, orange, yellow and green
Purple, blue and yes, all the rest
And though these words might sound mean
Colours are the best
For colours don't deceive
Colors don't reject
In order to believe
We do wind up suspect
To whatever they want
Yeah whatever they need
Whatever the outcome
Whatever the seed

Words are just arithmetic
And knowing how to spell
Like some crazy drug addict
Living in their hell...
Yeah just some brainwashed fans
Now living in filthy cans
Yeah the brainwashed souls, falling under some
Or another cult's control
Yeah just nasty brain spells, undertaken by some witch
Who just escaped
Yeah just escaped from hell

Colours,
Yeah colours never have, yeah never will drown
Never fall, yeah never fall down
They just keep us all within our ground
But always and always
Yeah they always do tell...

THE NATURE:

Winning isn't everything
In order to score
It takes a whole lot
Then a whole lot more
Psyche out the foe
Question, you will know
Learn out all you can
You must put on the show
Have you learned enough
To take a giant puff
On your better one
The one with all the stuff

You've run out of time
The time to act is now
Quench for thirst the lime
To repeatedly show us how
You are to play the game
Along, they're all the same
Some enjoy the fame
Pasted with their name
Oh my dear you lost

Well, isn't that the cost
I think I've had enough
Well, throw it all away
You're going through a phase
And on a perfect day
You'll work out the maze
The labyrinth within
I'm thinking I can win
When the time is right
You'll scream and dance right in
I'm in the middle now
The dancing of the set
Look and watch me how
I terminate the bet
The terminator's back
I'm back and just look out
For the nature of my sack
Then the nature of the spout
You see, you paid the cost
Of playing to win
You won but also lost
Now you know, you're in.

HEART STRINGS:

Here we are
On that page again
Here, are we
On that same road again
Running away
Running to the edge
Running towards our home again
Running forever and tomorrow
And for ever today

Hold on a minute
Hold on for a life time
Please hold on
To me again
Hold on to my heart
Yeah just hold on to me
As my heart strings have been played again

It was just our hard work
It was our heartwork
I can still see you
Still see your love
Behind your eyes
Yes behind those crystal eyes

My angel of truth
Please just remember what I say
No matter what "they" say
Just love yourself today
And it feels so good
So good
What else can I say

Don't let them put you down
You are good enough all around
Beautiful people are not easily found
Beautiful angels like you

Most are seen, but not heard
It doesn't matter anymore
Cause we have already settled that score
Already stopped, forgotten the lost
All forgotten
Don't let your mother
Or any other
Fool
Put you down
They are just what you said
From the start – all are just a fucking bore
And all the drugs in this world
Can't save them from themselves
And this I am sure
For now I have to go
But we will meet again
When you need me
Who am I?
Just call me
I think you already know?
You am I?
I really think so?
What is your life? – I want to know
Because you give me faith
Yes, you give me faith
In this world once again

THE BRUSH:

We started off unannounced
Made gradual progressions
We lived and travelled
At countless destinations
Though we live in the subconscious
Our minds travel far and wide
At times, we can't discover
The soul reasons why we hide
Experience, love and freedom
Is what we're searching for
Even though we find it
There still remains a door
To life there is a catch
To things unseen or heard of before
But if they make a match
We seem to yearn for more

Take life as it comes
Accept your limitations
Be careful of the bounds
Of countless destinations
Even though we're intelligent
We still lack time for living
At times we can be careless
And misjudge in what we're giving
Be still my restless heart
And chase away the blues
It's time to re-enact
Perspectives of your views
Again we paint a picture
And try not to regret
The things that you have painted
The brush cannot forget…

HIGH OR LOW?

Some still dreaming
Some still soothing
Yeah that's okay
Today still some peeling
Yeah some reeling, some plastic really
Yeah so pathetic
Really
Can we say

Are you
Are you the one still following
Yeah you are the one that kept swallowing that
Amaurotic faith
Hook, line, fool and sinker
And I just got accused
Yeah I got amused
I just was, was I?
Yeah always a thinker
And justice for all
Remember

Yeah remember that last cold December
You never couldn't
Remember
You can't sedate
Yeah you can't sedate all the things you hate

That's OK, always OK by me
That you still like to debate
Yeah still like to hate
Cause I am we

And we are many
Yeah always happy - deciding on our own fate

When I might
Yeah when I die
Don't really care if you cry
Yeah don't be afraid, but
Just put me and my music
In our own motorcade
Yeah in my own private parade
And make sure I'm in the shade
Yeah not near your zone
Cause I've always liked
Yeah always liked to be alone

Good it is
Now that your eyes really can see
High and low
Yeah above, behind and in front of you,
Even me
And maybe more
Let's see
Yeah do you now agree
That the darkness inside your fucking heart is not
And never was me
They always were your seeds
Yeah your foul seeds to sow

If you don't dance, you're dead
If you don't regret, you cry
If you don't live, you slide

Yeah if you don't love, U might just die
Time is a nuisance to all
Yeah to all that consume it
Like a slow releasing poison
Making it's mark
Sometimes making it spark in the dark

Yeah it creeps up
It creeps up on thee
To one day set you free
Yeah all in all sometimes we can see
The timing
And the tuning of our musical strings
Yeah the timing to set us free
A time to decide what to bring

Why does it exist
There is no why
It just is
Just our frequencies
Our vibrations
Yeah it simply does
In the sky, land or water
Space and time too
High or low - what's the time now
Cause I got to go
Yeah I've gotta go back to my own
Space time,
My time so I can glow
Yes flow down the river of my next life
Yeah, just say on with the show.

1 CLUE:

I've seen some life
Yeah, I've tripped the light, horrific
The drugs, chemicals and warfare
It's a little scientific
People, places and types of hair
Yeah, people and faces and many different strings
We sing the dance
Find sweet romance
Out of sight
And when I hear the blues
Yet, once again, it brings
Yeah, it'll be on the news
Of my strange end
Let there be laughter
Let there be joy

Yeah seek what you're after
And play your music
Yeah play the toy
One life
One heart
One chance, so entrance
Depending from which point
You take your view
Yeah you are to appoint the next dramatic clue

Play; life to the full
You're sure to find the one
Even if the fool is dancing in the sun
Flash, the lightning going through my mind

Yeah, quite frightening, happens every time

But what can I do
Yeah, just along for the ride
Even if to go away and hide
I'm climbing, scratching and searching
For anything at all
Even if I'm hoping for nothing at all
Or the initial call
Yeah escape the fall, just grab those strings
And run up the hill
Just one more wall
When DO they end: but I don't want them to
But maybe when I'm red in the face
Yeah or perhaps turning blue.

Always seek
Yeah always seek what you're after
And play with your music
Yeah play with that toy
One life
One heart
One chance, so there's your entrance
Depending on which point
Which point you take your view
Yeah you are to appoint the next dramatic clue
Yeah just point to my next one,

My one
And Only Clue...

ACT THREE
WHISPERS

NOVUS
ORDO
SECLORUM

371563 296790

THE BEAUTIFUL ONE:

She's the one
The one and only
The only one
To give me glory
She's the one
The crucial one
Make the way
She'll take all day
She dances
On a pool of air
Going places
From here to there
Turning suddenly
We take the wine
Tasting openly
It tastes divine
The cool wind thus
Comes in quite close
And majesty
Just seems the most
Breathless again

I wonder why
We don't say when
Just specify

She's the one
The magnificent one
Jumping wildly
Having fun
The one
The one and only
Heaven forbid
She seems quite holy
Let her stay
Another day
And then again
It's up to her
Why make fun
Of the beautiful one
Who sits and drinks
In the sun
Share with me
And be the one
Then stay with me
And we'll both run
Under the sun
We're having fun
Because in the end
There's only one

TIME TO GO

Memories in the past
How long can they last
I'm seeking today
I'm seeking tomorrow
Filling up my life with no sorrow

Forget about the pain
Out in the cold
The heat has now come
From that special someone
Now no more rain
And again she shields me
From the blistering cold

Using meditation in an open state
Time to go forward
Never to be awakened too late
Uncovering, yeah breaking the shell
It might only

Turn out thin - but you can't
Wait
To drink from your well
Suddenly standing
Standing up and looking
I will hold you close tonight
If that's alright
Yes drinking that cold, sweet and pale water
It's only ours to drink - already in sight

Life: it's ours to drink
Experiences quite unique
Remember them always
As it's time to go
And the next day we shall meet
It's time to go forward
Time to move forward and get off the street
Yeah it's our time to go forward.
About time to go forward

ALL THAT I NEEDED

I'm sorry for the pain
I'm sorry for my rain
Yeah I'm sorry for that dark rain
I finally found who I am
You never forgot who I really was
Always offering your help
I just took a long time
And finally found my true self
I'm sorry for the stress
Yes, I'm sorry for the distress
I wasn't just being selfish
Just a little stressed
Just a little distressed
That you would be
Taken advantage of
And fooled
By some snakes that never rest
I'll always care
Always know your true hearts
As I have from the start
In that time
And over time
Sometimes lazy
Sometimes
Not gentle enough for you

Still living in the fast lane
When it was the right time
To slow down
I would never lie to you
It was me causing your explosions
No fault of yours anywhere and
Still trying, supporting me

Still trying, not letting me go in your heart

But I was the one who was corroding
And should have and could have
Easily fixed it from the start
I guess I never got that part
Over time, you all will see
The real truth about me
I'm not worried anymore
As your own intelligence
Will let you see,
What you need to see
That I'm just a friend
With no evil agenda
Just a friend around the corner
Yeah it's still me
I'll always be there
Be there for my children

Anytime and anywhere
You know you can depend on me
See, it's still just me
Nothing here
Nothing said today is just a line
It's all about the truth
And in the end
That should be fine
All I needed was time

Time to heal
And time to deal
Like we all already know
All that I needed was time to feel.

Now all that has been said
That needs to be spoken
As you now know
Some things in me have always awoken
Yeah never really broken
When all is said and done
There is nothing lost
Yeah when all is said and done
There is
No love lost.

ONE MORE CLUE

I'm still developing a new thing
My present state of mind
Yeah, still searching
For something new to find
A change is good as a holiday
Well, I'm back into phase
This day, I'm already back
LIFE?, still like entering a maze

I'm always entering a different
Interesting experience
Yeah now...I know what it is
But it's still making quite a difference
To my unregretful life

The life I choose to live
My shape or your viewpoint
Make me still want to give
A certain shade of light
Allows you to see within
One that seems alright
But you never can win

I know you didn't know that
I still illuminate my life
So you can still
Only find
The scratches on the surface
Are a deeper, stranger kind
Reality can be so unreal
Did they ask you that

But please won't you tell me
How did it make you feel
In a given point of time

I might be something else
Not somewhere or with someone
And not even a mouse

It's sometimes clear and oblique
Yet, that's a state of mind
Still unique
If it's something new I find
I'm in a later stage
Change is always my friend
No matter what the question
With me 'til the end.

My colour hasn't changed
My face still looks the same
So why are some of you
Playing a different game
Rules and regulations
They're all the same to me
They're mind manipulations
To one day, set you free
Find the answer to the question
The one that fits right in
Make it act like the solution
Then the answer to the sin...

Come drink and dine with me
Intoxicate within
We'll have a feast or two
Yeah, and see if we can spin
Understand adventure
Explore to the unknown
Pioneer your own life
Freedom is yours
Yeah freedom is yours to own...

A DIFFERENT WAY

There's only way to go
By sea or air, or by snow
Whichever way, you'll never know
The only way, you are to go
Travel far, you travel near
Take heed you are by far
The gamest one to not know fear
No matter who or where you are

Be still my silent pet
I love you, you know that
The cool night draws in quickly
Draws and breaks, forever sickly
The candle burns out to an end
The creature looks with open eyes
What messages are we to send
Communicate with pseudo lies
Run and fail, yet once again
Who speaks with an open tongue
Fall, get up, and then
His neck needs to be wrung
Ahh, I'm back in sync

With this I'll have a drink
Sit enjoy, relax right back
Eat, smoke and try to think

Deserve to thus repeat
Let's ask her while she's running
She's very agile on her feet
Deceptive and quite cunning
Unusual for this time of year
A little early, or is it late
By the round curves on her rear
Can I afford to hesitate
Duty calls, I must report
Get back to empty minds
Overseas, I will resort
To abstract, strange, delectable kinds

Hey wake up, it's time to go
Go and find the treasured land
Grab the dream, go with the flow
Come back and hold it in your hand
Confusion flies, let it grind
Of which path I am to take
Maybe the one here in my mind
Whichever one, for heaven's sake
Fate be the judge of us all
Whichever way, a different way?
Fast or slow
The way to life, big or small
There's only one way to go....

HOME RUN

Did you hear that thunder scream?
Everything is not what it seems
Did you see that lightning strike
Yes right on top of you
It just missed me too.

Stars falling down
Your empathy starts to drown you, please!
Don't throw away those tears
Yeah just don't throw away those years
Don't throw away these years
Always watching
Always listening to the skies above
Not many people –
Look up these days
Not even a star gaze- far away
Too busy following each other
Around in their own little daze

Just say you'll be home soon
Very soon, I bet
Yeah home very soon, even
Better if it' so on the moon
I know it's not April Fools anymore
Sorry about that
I just had to settle another
Personal score
Just like a home run

I think you understand
Cause this is just the beginning
Not my last stand
It's my home too
I'll be home soon
Yes, it's my own home RUN
We heard the thunder scream
Yeah everything is not
What it seems
Most of the world is a stage
But not anymore – not on my page
Did you see the lightning
Before it missed me?
Did you feel the lightning
Enter your body
Yeah it only sparked you
Sparked you today
But I'm on my home run every day

Always watching
Always listening
Yes to the space above
Don't worry you won't be going
Home soon, not even to the moon.
Just find your own home run
Maybe then
You could reach your sun

SO CAN YOU

Shakespeare did it
Plato too
Everybody
Wants in on the act
Dramatise
Specialise
Anything
To be exact
Conflict within
To find yourself
Anything's worth it
Find the elf
If they can do it
I can too
And if I can do it
So can you
Remember
They're not obsolete
In November
I'm complete
Try December
Now it's a better month
For in forever
We take another lump
Choose to recall
Engross her now
Catch the ball
I'll show you how
No, if they can do it

I can too
And if I can do it
So can you...
NO END:

I am resting in this cool lagoon
Some stars that live forever
How I want to touch the moon
For it will someday sever
This strange pool's like a jewel
White lightning wants to hop
If you swim, you're not a fool
In this misty mountain drop
Fresh clean water, cool as ice
The wonders I perceive
Like a pseudo paradise
Her beauty does deceive
The water here's divine
Nature's so mysterious
Beauty surrounded by crap
Deceiving eyes, yet so serious
Devouring like a Venus Fly Trap
Some turns we take in life
Some, more of a natural cause
We take them for the better
And waste them, just because
Your life is only tripping
And don't you ever pretend
As there's no beginning
There definitely, is no end...

TIMING THE MIND

I can't imagine the time it takes
To amuse you
And I won't waste my time
To just confuse you
I can only tell you the time
That I will confess
You don't seem to look
Don't seem to listen
You just seem distressed
Do you consent?

Brick walls don't bother me
Or neither should they you
However your head space
Doesn't make any sense

How can I see
How can I find the person
You really are
When you just pretend you
Are a walking charade?
Are you a walking clique?

I can try
But, it's going to be
Very far away
If you're not free - now
From all your hating
Thoughts?
Can I trust you with those talons?

Remember a dragon
Can turn – turn at will, even
Into a butterfly which
Also bite back sometimes

Sometimes kill
How do I know
That you will not try to bite?
How do I know? – I never forget
How do I know?
You will leave me alone?
Well I haven't even sung
That song yet

Just look after your own
Life and ego
I'll happily get on with mine
I'll be just fine – in time
Yeah fine in time
Just not standing
On your crooked line

Got the butterfly's
In your stomach?
Found the cockroaches
In your mind?
Just watch out for your delusions and self-denial
They really won't
Kill any of your time

Is it time for you to have
A break too?
Cause I aint the fool
I aint the joker
I'm just in full view
Just a helping hand
Got the time on my side
Think I've got the right time too
The right time beside
The right time
Behind my mind.

NEED A HAND

Be there pleasure
Be there treasure
Find the way
And make the measure
Tell it to me straight
So I won't be late
I don't know if I ate
The words to hesitate
Words of wisdom
Words of song
Say them all
Then sing along
Seek perfection
You're sure to find
The true connection
To body and mind

Heal them all
They're all in fear
Big and small
Both far and near
Then set them free
Their time to leave
Next to thee
I do perceive
Pull me up
I need a hand
To try and help
To understand
Believe what I say
And don't be late
For then I may
Soon hesitate…

WHISPERS

Cool as a whispered prayer
I'm over here
And you're over there
Playing with mortality
You're dancing on fire
On a very high wire
The echoes are heard
From someone or something
Maybe a bird
An injured wing
It strives to fly
To an inviting plum tree
So it can eat
And rest peacefully.
Behold what a sight
The day turns into night
For a few seconds only
The moon hides and passes slowly
On their way
The men witness this
Wait for the call
To ominously play
They dream another place and time

You'll have to pray
The hourglass has no sand,
The sun gets close.
Time has run out
Where did it ll go?
Poured through a spout
It went far too slow
What a shame
Time wasted and
All over the land
Omega Day..........

REASONS

Sometimes we cry

Yeah, sometimes I cry
Then I look to the sky
And hear how it chimes
Yeah see how it shines
Just think in the now
REASONS for us right NOW

REASONS to stand up AND fight -
Don't hold back
On all the 'Insurance' Files
Yeah just remember the right ring tones and strings
Yeah the right number dials

In our beginning, we were just singing
Yeah this was never our world
And I didn't mean to
Take your angel away
Don't blame yourself
Yeah don't hurt yourself
To make everyone pay
Just to make everybody pay
Yeah, no other reasons
But just do it our own way

This was never my world
I know you didn't plan
On taking
Your angel away
But I won't give up, never give in
Yeah don't kill yourself for any reason
Just to make everybody pay
Just to make

Them all pay
Now we're no longer blind
No longer blind
Yeah even this time
I know we'll be just fine
It's just our turn to shine
Never be blind
Never be left behind
Yeah never be left behind again

The calm is still coming
But there is no more
Yeah there is no more running away
Yeah, I'm not a hero
Still just a simple human today

So I'm calling you all
Calling you all
Yeah those good souls
Don't you
Yeah don't you ever
Give up on your life
Cause this is far
From over
You'll always have strong shoulders
To bear all the heavy
Weights
Cause I want to live
Live my life
Yeah, no more lies,
Yeah no more cries
Just want to live
Yeah just want to live the right life
Again
Yeah, no other reasons

CROCODILE CRIES

Are you hurting?
Why are you crying?
Again? Is this real?
Is it just me
Dreaming?

I'd rather be crashed out sleeping
If that's your headcase tripping
Foulness slipping
Yeah falling out of your mouth again

I'd rather be crashed out dreaming
If this is one of
Your reptile games
Then I could find
Another race
In a better place
Yes with better race
Where there is no more lames

Just take a minute to rest
Like the rest of us
Take a minute to ponder
About the outer world stage
Without increasing your stupid rage

Your crocodile cries don't scare me
Crocodile tears don't confuse me
Yes
No confusion here you see
Yeah no confusion for me
Only for you
And
You can keep your crocodile tears
Forever
For you
All alone

You're welcome to go and
Pack your bags
Then shed some skin
I cant do it
For you
As I'm not reptilian
No reptiles even in my bin

I'm sure I only cry when I need to
But nothing like you
I'm nothing like you cause all
You are
Now is
A very deep scar

I'm not sad now
Don't know if I was
Even hurting then
Doesn't matter anyway
Won't be seeing you in this
Life again

Am I crying? Hell no
Am I lying
No way, but never afraid
Never to myself
Even if it's the end
Cause I'd rather die sleeping
Crashed out dreaming
Than listening to all those
Crocodile cries again

Do you remember what I tried to tell you?
If not just remember
I tried to help you
Now it's too late
Now it's time to say goodnight
Little crocodile, your teeth are gone
Yeah no more bite
Cause you've just lost it all
Lost your last, everlasting fight.....

BURNING

Looking out for love
Just be careful
When you turn that key
And don't turn
Don't turn the knob if
You think it's going to
Really burn

Burning your hands
Are you burning
On your own front door?
Yeah burning my hands
No I'm not playing
No I'm not joking
I think someone's
Trying to settle a score

ARE YOU SURE YOUR KEY
NUMBER IS
REALLY 333?

Absolute Power
Can lead to absolute betrayal
Now, I'm not laughing
But be careful
When you're still a fool
Well it's still me here trying
Trying to help
Help you with your right tool
Yes it's just me who is

Trying to stay cool
And yet,
I'm not even close to you

Do your dreams always work out
Even in another time?
Why don't you just
Keep drinking
Keep drinking your own pathetic stale wine

I'm not playing, I'm not shaking
With you any more
I'm not going down
Yeah I'm not going to burn for any score
Haven't we seen all this before
Behind your door?
Yeah, haven't we seen it all before

Now look at your hands
Touch your face
Yeah look at your own eyes
Those eyes do tell
Now tell me who's burning

I've already tried cooling you down
But yours and even my
Wishing well
Has already
Gone
Gone bone dead dry

Yeah that's why you haven't heard
You own
Alarm bell

Now you're burning
Yeah now burning
Burning in your empty cell
Burning yourself all
The way to your own hell

333 Is it just a number?
Some choose it for this
Some choose it for that
People living in certain realities
Already know that shit
Our people don't worry
Yeah we just laugh at your ignorance,
It's just your own deluded fit

No worries
Yeah no concerns about little fire ants
Like you
Maybe only a select few
Yeah maybe some of those
Little fire ants
Will
Never see
Never fe Free
Or never enter
The only one right cue

ONE LAST ROUND

One more fight?
I aint surprised
One more fight?
Don't play it down
In front of me
One more fight?
Who's your next target?
Well just tell me
Now who's your enemy?

Now you're choking
No, not controlling
Just so hypocritical
Yes your own pathetic
Egotistical nightmare
Go back
Go back to your silly little book
And stare
Yeah stare at your
Big lie
In the sky
Now by a telescope
Look for truth
Look for the real creation
In the same sky

One more round
Are you kidding me
One more round
You're listening from the ground
One last round?
I'm not your enemy

Everybody's sick

Sick of your lies
Everyone's tired
Tired of caring, tired of your bullshit
Tired of your reptile cries
Can't you see it
See it in their eyes

Just take care of yourself
Remember snakes get bitten
Every so often
Yes snakes get bitten
Snakes get bitten too
Snakes like you

We see them surfacing
Easily I should add
Popping up
Out of the ground
Every now and then
Just remember
Snakes get stepped on
Snakes get stomped on too

Is your blood cold
Or is it burning
Inside your veins
Cause here we go again
Here we are, again in your head
In your head again

I'm not to blame
When your world fades to black
I'm not the one
Who chose to start on you, you hack

Are you insane?
You've done all
Tricks like that by yourself
Even after that evil feeling attacked
So take your eyes off the trigger
Or your eyes will start to blister
Here we are again

So tell me now, how
Does it feel when the poles are reversed?
When the poles are reversed?
Negative and positive
Swirling inside your brain
Now it's just up to you
Up to you on how you train

Have you got a fever? cold sweats,
Hallucinations,
No?
I think its just grand delusions
Just see the truth doctor
And get the right pill
Or you just might have to
Stay on the upside
Yeah, the wrong side of this hill

Still don't wanna stand beside you
Don't need you to hold my hand
Thanks anyway
But enjoy your fakery
Enjoy your own delusion
All that's not for me
Cause I live on the ground
Not in the clouds

Yeah it's not for me

One more round? – I'm glad it's over
One last round? – A bit one sided
If you ask me
One last round?
You were never a real threat
Never considered an adversary

One last round?
Couldn't be bothered
Just a waste of time and energy
Your plight was over before it began
You never had a chance from the start
This is why you ran
Now it's time to say goodnight
Time to say goodbye
Right from my heart

HOUSE IS DOWN

Why did you play with my fire?
Yes tell me why
Did you set me alight?
I never wanted to fight
Don't shoot the messenger
That's an old mistake
But you've given
More than I can take

Why did you play with your fire?
Yeah tell me why did you set yourself alight?
I'm out of water
All out of that type of water
Now you're going to burn
And you can't even swim yet
Forget about me
Because I've forgotten about you

I have many friends
Even in a dark storm
Yes, right through the storm
Don't need your words
Don't need your care
I don't even need your stare
Just don't need you at all
Because you're about to fall

I'm going to
Bring the whole thing down
Your whole system down
Yeah I'm bringing
The whole house down

The whole, yeah even the burning flag
Yeah bring this whole dirty house down
I didn't start this war,
But it's already ended
For you should
Have opened your eyes
As you didn't even know it
Know that
It was always in our bag

We ain't scared of any truth
We ain't scared of your police
Ain't scared of your media
Yeah we're just not scared
Of monkeys
Or any dark charades
Yeah puppets like you - even if you're in the sun
Watch the dominoes fall
One by one
Until they destroy your fun.

We all gave you a chance
A chance to back away
But no real listening
On your end
I'm afraid, so I'm
Gunna bring the whole lot down
Today, not tomorrow
The house is down

I'm not scared of your drugs
I'm not scared of your hugs

Not scared of your head
Nothing left in there
Nothing left to be said
Not scared of anything
'Cause I'm free
And always have been
You just got no fucking idea
What I've seen
We are unbreakable you see?
Are you a puppet?
Are you a muppet?
Which one is t 'Cause
I can still see those skeletons
Falling out of your closet
Even out of your drawer
While the monkey is still latched
Tight, watch out for his bite
Latched right on to your back
And I bet you, - you are sore

The house is down
Your house is down
Which one is it?
Cause I don't really care if
I'm not a hit

"Cause this isn't music
And I'm not a band
Just five middle fingers
On my motherfucking hand"
Just like Brian Warner said
Yeah just like Marilyn Manson said
Yeah rue that!

PERSONAL FLATLINE

Sun Tzu said it best
Put your enemy's through
Their own personal test
I won't be deceived again
Yeah I won't be burnt again
Yes I'm still wearing my invisible
Bullshit proof vest

I can see your scars
See your holes appearing
So deceiving yet – some are moving
My ears nearly got hooked
Hooked on your tongue claws again

Life is like a game of chess
You haven't learned that?
And you're an old school VET?
Stop playing checkers
And get something off your chest –
Understand it yet?

My mind won't be switched off
So don't try and treat me like your bitch
I love sniffing out the rats

And stomping on the snakes
No. no more mistakes
I won't listen to any deception
Won't be fooled by your eyes
Cause everybody, already knows
When they see
A wolf in disguise
Oh yes I can see
When you're smiling back at me
If you like psychological warfare
Go play with the easy targets
Like the sharks in the sea.
If you try it on me
I would just show you
Your same reflection
Yes your shame reflection –
In your now blind eyes

Play with your own deception
Go play with your own kind
Yeah not mine
As there's nothing here anymore
That you could
Even find

Look in the mirror
You might find it pretty dark
Might find it cold
Come on you're pretty bold
But before you break the mirror
You need to see –
See your black shadow smiling back at me

Don't come running for a light
Your own black shadow even just died
Don't come running to me for a spark
As I don't do favors for fakes in the dark
See that, feel that? Smell that? – Now you're on fire
Now you're on – your right line

I told you – that's what Hell's really about
Little man, now you're lying down
Dying with your head in the can
With your emotional flat line
A personal flat line
Yeah now you own a personal flatLine.

NUMB:

I Got The Feeling
That Feeling Just The Other Day
Went For A Walk
Couldn't Hear Any Talk
Yeah Couldn't Hear The Normal
World Talk

The Light Seemed Dim
And No Shade Could I See
Just Darkness Looking
Looking Down On Me

When I Looked Up
At The Sun
It Didn't Look Happy This Day
Doesn't Look Happy Today

Yes Still Blinded
Blinded A Little
But I Couldn't Look Away
Couldn't Move My Eyes
Just Couldn't Look Any Other Way

Light Just Got Darker

Sun Still Didn't Look Happy
And I Was Just Distracted
Yes We'll Were So Detracted
When?
Then
Our Sun Just Turned Black
Now Just A Black Star Shining
Sending A Darkness
Like A Real
Deadly Night Shade
Over Everyone
Killing Our Fun And Killing Our Sun
As We Fade...And Many Asking Why?
But

We Were Too Dumb To Run
And Too Dead To Die
Yeah We Were Too Numb To Run
And Too Dead To Die

When Some Of Us Woke
Mona Lisa's Face Was Melted
Some Still Alive
Many Just Cremated

But Most Of Us Still Very Elated
Yeah Just Sedated

Millions Crying
Asking What They Did
Asking What We Created

Now Just Darkness
No More Morning
Yes No More Dawn
But We're Going To Find A Way
Find Another Way
To Spawn
As There's Always Another Way
To Just Mourn
Now Is It Time
To Just Spawn...And Not Ask Why?
Cause:

We Were To Dumb To Run
And Too Dead To Die
Just Too Numb To Run.....And Too Dead To Die..

STRIKE 3 FOR THE COBRA?

Still Thinking
Still Moving – Getting Very Close
I'm Not Worrried
Just Waiting
Waiting With My Noose
The Cobra Is Cunning
But I'm Just Too Quick
And The One With
The Bigger Stick
This Cobra Thinks
Thinks He Is king
But I'm Not Part Of His Kingdom
Nor Part Of His System
Just Death I Will Bring Him.
Keeping One Step Ahead
Always Is My Game
Only Showing My Hand When the Cobra
Goes Insane
The Art Of War – I know You Understand
We Have To Play First and Last
In This Stupid Game
Be Careful Cobra, cause
I'm The Black Mamba – Death Incarnate
I'll Shut Your Mouth
Swallow You Whole
Yeah Swallow Your Soul
Watch Out Cobra
I Think I'm Spitting
My Venom now
Watch out Cobra, You Cant Be Saved
Looks Like
You're Digging
Digging Your Own Grave
I Will Spit My Venom

Spit The Venom In Your Face
Just Spit On Your Grave Again
Now You Know How
It Tastes
Now You Know How It Hates
Cobras Don't Seem
To Like Acuphase
Nor Do All The Monkeys And Rats
Black Mambas Just Spit It Right Out
Right Back In Your Dark Faces
Yeah Back Into Your Dark Places
No Effect
Yeah No Effect On The True
No Effect On The Strong
Yes I'm Still Thinking Cobra
Moving
Now I'm Getting Close
Keeping Ahead Of Time
Patience – That's My Skill
This Black Mamba Soon
Will Go For The Kill
Goodbye Little Cobra
Never See You, Never A Next Time
You're Lost Inside, Shedding
Shedding Your Skin – Not Mine
If You Ever Find Your Way
Outside Again
I'll Be Waiting – And Will Suit Just Fine
As I Will Always
Have The Time
The Time To Find You
And The Time To Remind You – That You're Already
Dead
Did You Hear What I Said?

'AMERISTRALIA' : THE 'Main' Man

There Is No More Room In A Seedy Bar
Yeah No More Room For The Monkeys, Muppets,
Rats And Snakes
All Have Made Their Last Big & Evil Mistakes
Yeah Now I Just Laugh, As It's Living In The Now
Some People Still Want, Still Need Money
Yeah Still Some Needed So We Can All Eat And The
Rest

But Even Our Taxes Are Are Still An Ultimate Test
Our Ultimate Enemy Thinks He's Safe In Some Sort
Of Nest
Just Forget The Banks
Just Bury Your Money In Your Own Tank

(CHORUS}

Some Reptiles Still Even Attack
Yeah Even When Their On Their Own Backs
Double Crossing Each Other At The Word Go
Just Now They Can't
None Of Them Can Catch Me
All Of These Inside Moles
Will Soon Be Sealed Forever
Into Their Dark Holes

Yeah Into Their Graves And Never Be Saved
Ameristralia? - That Modern Idea, That Modern
Thought
Yeah That Modern Term Isn't Recent
That Deception Started Way Back During World

War 2
17 Years Ago
1996, Yeah 1996, My Eyes Were Opened Again
At 22 Years Old
Yeah Others Too, Much More, Not Just Me,
Just The Young And The Free
And No One Can Break It From Us Or Me
Yeah Can't Break It From Me
Yeah All Of Them Now Really Buzzed And
Scattered - Who Cares, It Really Docsn't Matter
Cause They Are All About To Break
..And Shatter

(END CHORUS}

They Always Were, Still Now Far Behind,
Yeah Never In Front, Except To The Blind
We Are All Warriors
The Soldiers Of Truth
Yeah People Like Us…
You Aint Seen Nothing Yet
Everyone's Different, Nothings The Same
But Some Just Behind Certain Starting Lines
Others Just Behind My Starting Line
But That Is Also Just Fine
Cause I Have The Right Time
The Right Time In My Heart, Head And Soul
It's Just About Real Self-Control
Everything Has A Different String
Just My Dragons, They'll Always Win
Yeah Just Feel The Right Dragon And He'll Redirect

Yeah He'll Feel And Connect With You And Your
Friends

Maybe The Green Fairy Has Her Reasons,
But In Moderation It Keeps Me Motivated And
Dedicated, Just Like Truth And Art Together
Never Too Elevated, Never Too High - Just
Yeah The Right Balance
Just Another Libra Tipping The Scale
Just Watching Another 4 Eyed Snake
-
The Small, Yeah the Weak
Little Dirty Rat
Johnny
Howard Hiding His Crime, But He's Near The End
Of His Timeline

Any New Enemies, Just Pass And Turn Them Around
Just Drive Past, They Can't Catch You
Yeah, Do They Still Have The Mountain?
Not Far From Me Or You
Mt Evelyn Hides A Lot Of Lies...
And More Truth Overall...
Even Before I Was Born
But That's The Nature Of A Real Spawn
That's Just Between Me And You

Can't Catch Anyone As Now They Are Blind
Always Come Last, They're Not That Fast
They Can Travel If They Like With Their Sword,
Gun Or Bat
But They Can't Kill Me, Cause I'm No F#%king Rat
Just Give Them A REAL Heart Attack
Yeah Just Give Them A Real Wake-Up Call
As In The End They Just Fall And Fail
Then For Eternity They Will Wail

Yeah They Never Swallow Real Water Again
Then Have Their Own Personal Flat Line In Time
We're Not Scared
We're Not Prophets
Just People With Deep Pockets And Using
The Right Tools
Yeah The Right Sockets
Yeah We're Just People With A Good Memory Bank
Memories From A Decent Hard Drive Selection
A Special Collection Of The Real Truth And Real
Lies, Maybe Even Some Innocent Cries

(REPEAT CHORUS}

There Is No Guilt Hanging Over Me
Yeah No More Guilt Left On My Life Tree
So Know, So Now I Can See Through All The
Corrupt And Control Freaks
On The Street And In The Air
Yeah, Even Some Hiding Well In Their Own Secure
Lair

If You're A True Realist
Don't Be Just Another Fucking Troll
Rising Or Slipping Down The Corruption Stink Hole
Feel Free To Whisper, Feel Free To Gossip
Yeah Feel Free To Lie
I Don't Care If You Fall
I Don't Care At All
Gave Enough Warnings

Especially Your High
Kill Ratio
Most Didnt Count
As It Turned Out - It Was You And Only You
Yeah Your Crocodile Tears
Crocodile Cries

Yeah Your Fucking Deluded Lies

Even Eagles Can See You
From The Crimes Above And Far
Yeah Even A Dove Can Make You
Can See You For The Fake Fucker You Are
'Cause You Can't Outrun Everyone, Anything
Can't Outrun Me Even In Your Car
Knowledge And Imagination
None Of Us Are Just Fools
Already Behind You
Yeah Also Still In Front Of You
You Never Had And Still Don't Have The Right
Tools

(Repeat Chorus}

Just Like A True Memory In Λ Strip Bar
Life Long Stars, So Close Sometimes
That Special Part Of The Brain
Sometimes Has The Right Effect
On Others You Find
But Just Be Careful Not To Make Your Own Or Their
Brain Begin To
Swell To Much - Within Life's Big Well
Yeah We're All At Risk Sometimes
But Now We're Ready, No Risk At All
We've Expressed Your Will
, Not You, You Never Had Control,
Just Like Another Internet Troll
Living Under Their Own Dirty Bridge Of Deceit And
Medical Scripts Piling Up At Your Feet
Never Had Any Trust, Just Like 2 F#%%ed Up
B$%#hes Who Thought They Were Too Smart
Betrayed All Your Own Trust
Nothing Left
I'm Not Giving You Any Water

I'm All Out Of The Special Brand
Yeah, I'm Out
I'm All Out - Stepping It Up A Gear
My Water Is For Kids, True Friends And Me
Because We Can Already SEE
I'll Never Shout Out - Too Soft
What's The Point
Just Tell The Truth
Yeah Never Fake, No Fantasy Shit
Even If You Still Want To Try And Lie
Yeah Try To Attack Within
No One Will Ever Have Your Back
And I'll Always Win
We Know Who Our True Friends Are
Even From A World Apart
Yeah Even From A Far
Maybe Even From Another Star
Maybe One Day I'll Need A Real Fast Car
Yeah, Just Like My Purple
R33 GTR

A DIFFERENT DAY

Tell Me Of A Different Way
How I'm Supposed To Say
How Are You, Oh G'day
On This Clear, Bright, Fresh, New Day
Well, Isn't That The Way
You Know Just What To Say
Here Where I Do Lay
I'll Scratch Off All The Hay
Maybe Light A Jay
Then Go All The Way

Tell Me Of A Different Day
If You're Scared, You Will Not Say
This Is A Brand New Day…

ACT FOUR
TUNE YOUR SONG:

TUNE YOUR SONG:

Peer In Your Place
See All The Waste
Learn To Conform
Not To Reform
And Without Haste
Without Our Faults
We Cannot See
What Is 'Wrong'
What Is Meant To Be
But Who's To Search
And Find The Clear
One Day I Might
Soon Disappear

I'll Go Ahead
See Worse Instead
And Better Yet
I'll Find My Head
I Think
Who's Out There
Where's The Freedom
Most Everywhere
Can You See It
Forget The Head
My Soul, I Don't Know
Where's The Red

I'm Sure To Go
Get Help Soon
Learn To Find
The Golden Moon
And Not Your Mind

For We Are Placed
In Hideaways
Then Escape
To Brand New Days
Never The Case
I Cannot Go
Just In Case
They Are To Show
Hymn Along
Dance Very Quietly
Tune Your Song
And Keep It Partly
Smooth And Rough
Civilized
Had Enough
I'm Hypnotized
Hey, Look Out
I'm On The Way
Onto Where
The Brand New Day...

JUST BE ONE:

I'm Still Linking Places With Events
We Notice Faces
Yeah In Desolate Descents
I See The Shadows Running Below
Yeah, An Abandoned Paradise
...And How I Want To Go
Here Is A Liquid That Tickles Your Mind
Colourful Places Of No Other Kind

Yeah Louder
Louder Footsteps Running Wild
Through The Corridors
Like A Frantic Child
Seeking The Passage To A Greater Domain
Yeah Seeking The Wisdom, And Play The Game
Be The One, The One And Only
The Only One To Seek The Glory

Does It Exist, I Really Wonder
Or Does It Exhale The Wondrous Mist
Yeah Search Your Domain And Seek Your Future
Find The One Who Inhabits Your Gain

Let Life Play, Its Ugly Game
Yeah Let It Describe Its Unique Name

Dreams They Come And Go
And Sub-Consciously We Think We Know
Yeah Look A Shadow, What Can We Find
On This Dark Topic In Our Intricate Mind
Yeah Just Remember
Remember To Sometimes Be Kind...And Never Go
Blind

Yeah Louder
Louder Footsteps Running Wild Through All
Corridors
Like A Frantic Child
Still Seeking The Passage To The Greater Domain
Yeah Still Seeking More Wisdom While Playing This
Game
Just Be Careful Not To Go Insane
Yeah Be The One, The One And Only
The Only One To Seek The Glory
Yeah The Only One To Share That Story

MAKE A MOVE

It's Burning To A Tee
The Flame Has Set Alight
The Rising Of The Free
Smoke And Outer Sight
I Need It For The Glass?
I Need It For Myself?
I Need It, Is That A Fact?
I Definitely Need To Be Exact

It's Oozing Down My Throat
It Starts To Hit The Sweet
Spot The Other Goat
I'm Terrible To Meet
It's Time To Down One More
Give Me Time To Score
Now I'm Getting Sore
It's Time To Make A Move
I Wonder What I'll Prove
Feed The Bag To Me
Feed It Endlessly
Saliva Needs A Break
Time To Have A Drink
To Intoxicate
And Reach The Other Side
But Isn't That A Joke
You Are Such A Fool
Now You Start To Choke
Enigmatic Smoke

Is Filling Up The Room
Wait A Minute, Boom
Get Closer With The Zoom

Let's Fill Another Room
Let's Cut Up Some More
See If We Can Score
Come On And Have Some Fun
Dance Up On The Sun
Oh, I Think It's Blocked
Just Leave It To Me
I'll Unlock The Locked
And Tend To Set It Free

Exhale The Inner Trip
It's Going Out Of Me
I'm Starting To Rip
Out Places Near And Far
No Matter Who You Are
I'm Starting To Regret
The Loss Of The Good Set
Unzip The Zipped
Oh Fuck, Hang On
I'm Ripped.

Logical?

Strike Another, Poke Another
Yeah, See That Lightning Flash

Then Discover
Dramatic Slash
Yeah Or A Dramatic Stash
It's Creeping Up
Up And Up
Which, The Way
There's Only Up
Then Comes Down
Logical, With A Frown

That's Technical By The Way
Physics Rules And Sucks
Let's Just Say
Run The Mucks
Now It's Getting Stronger
Little By Little
Longer And Longer
And You'll Realise How Quick It Is
With Surprise, The Unknown Fiz
That's Technical, Commercial And All The Rest
Logical?
Now That's The Best

In The West Take A Rest
Cause In The West
Find The Best
The Best Of What, Only You
Know What It Is...And Break Me From
This Burning Truth Fiz...

SING:

I Was Wrong To Go On
And I Was Right To Be Right
What Ever The Case, Where's The Song
I Plan To Sing And Sing All Night
I Was Right Last Night
An d I Was Wrong To Be Wrong
What Ever The Case, Who's In The Right
In Choosing Someone Else's Song
Can I Go On
Why's The Cat Lying Flat
Who IS That
Near Your cat
It's My Cat O.K.
You Are A Little Brat
So Don't Spend All Day
Looking For Your Cat
Your Cat & Mine As Well
Are Drinking Sweet, Warm Milk
Dressing Near The Bell
With The Brand New Silk
I Want To Play Forever
For ever Yesterday
Dance And Play Together
And See If We Can Stay
Together And In Sync
Come On, Lets Drink This Drink
And See If We Can Sing
Hey, We'll Sing Anything...

HOLIDAYS:

It's All Happening
I'm About To Go Ballistic
It's Getting Real Close Now
And A Little Plastic
I'm Approaching The Finale
All Systems: GO
Tell Me, Are You Coming
How I Want To Know
The Long Road I Have Travelled
Is Very Far, But Few
And Although I Have Cancelled
The Long Awaited Due

I'll Go Anyway
No Matter If They'll Stay
For On These Special Yeah These Perfect Islands
With The Perfect Day
Are Women All Around
Unescapable Ground
Wearing Next To Nothing
And Listening To The Sounds

Nothings Better Than Something
Well, Isn't That The Truth
Clothes; Here A Waste Of Time
Where Everyone's Half Naked
Holidays, About Damn Time
Yeah, I'm Still Drinking, That Succulent Wine..

MY PATH:

Now I Have Found Myself
Now I Know The Truth
It's Easy To Weed
Yes Weed Out The Lies
Still Weeding Out The Traitors
Still Witnessing Many Deceptions
Yes These Are True
My True Perceptions
We'll Be The Last Ones Standing
When We Get Out Of The Maze

Yes, We'll Be
The Last One's Standing
At The Front Of The Line
Empty Cities, Empty Skies
I Know, It Looks Amazing
So Quiet
So Weird
Just Extraordinary - Yes
So Why Don't You
Take A Chance On Me

We Don't Need Infamy
We Don't Need Politics
Don't Need Religion
Just Leave Us Be
Leave Us In Our Own Private Room
Yeah Private Room 11 – Number 1,2,3, Three and 3
6 & 9

My Path Has Been Found
Not Going Straight To The Top
My Path Is Long And Winding
Sometimes It Can Even Corrode
Yeah Taking Easy Steps
Easy Steps On This Road

Yes Our Path Is Now
Found The Path Right Through
My Mind
Yes Right Through My Soul
It's My Path And Now I'm In Control
I Just Found My Path
And You Were Always There
– Never Letting Go

I Know You'll Always
Keep Up With Me
It Was You That Guided Me Back
Back To See
When I Lost My Way
You Were There In Some Way
Every Day

It Feels Like Our
Hearts Still Beat The Same Drum
Yeah Don't Fall Down
No Falling Below
Like Others Before
I'll Always Find You Even If You're Torn

If You're Ever Lost
Doesn't Matter Where You Are, I'll Come For You
Life Has Just Begun For Us Two

You Don't Need To Worry Anymore
I Sense Your Love Every Day
But Remember, Angels Like You
Need Sleep To
I'll Always Be By Your Side
Even When The Dark Tides
Try And Follow You In
Yeah I'll Always Love You
Everyday, Still,
Even Today
As In My Heart, You've Always Been.

THE CHANGE:

If All The Pieces Are Gathered
And Put Together To Fit
The Puzzle Of Life That Remains
Is Dark And Cannot Be Lit
For Time Is Spent On Searching
The Things We Cannot Find
Time Is Lost And Then Forgotten
In The Labyrinth Of Our Mind
Curious In Our Nature
We Set Out To Explore
We Wonder How Things Happen
Always Wanting To
Yeah Always Wanting To Know More

No Stone On Earth Escapes
The Clutches Of Mans Greed
Earth, Where He Lives And Rapes
Is Devoured At Lightning Speed
Progress They Call It
Breeds Destruction And Confusion
Our Orbs Need Redirecting

To Enable Us Salvation
For We Have Wealth And Power
That Turn Us Into Robots
We Must Respect And Uphold
The Unjust Laws That Govern Us
Yet At Times They Are Waived
To Suit Those Who Are Evil
Corrupt, Influential And Capable
Of Controlling Our Worthless Lives

We Are Still Just Little Fish
In The Sea Of Damnation
Brought Into Without Choice
Of Another Civilization
That Never Hears
Yeah That Never Heard Our True Voice
Some Day The Change Might Come
But Until Then Pretend
That None Of It Exists
Yeah This I'm Sorry, To Say My Friends
Only To Find No End...

MAYBE:

Maybe I Was Wrong, To Go On
Really Wrong
To Go On
And On
I Was Right When I Was Right
And I'll Always Win That Fight
All I Really Did Was Plan To Sing
Yes, Sing All Night
Can I Go On? To See What I Bring?

Maybe I Was Wrong, Not This Time
Not This Time At All
I've Always Had A Fine-Line
But Chose The Right Line
Not Picking On The Small
But Sometimes I Did Fall

Right Equals Right
Can Let That Rest?
Any Mistakes? – Yes, I Confess

What Ever The Occasion
No One Is Always Right
So If I'm Wrong Don't
Get Uptight
Because I Will Not Fight

Seeing Through Your Lies
Are A Little Easy For Me
I Came From The Same Neighbourhood
Remember – I Won That Fight For Good

Won From Your Very First Major Lie
I Was Next To You
At Your Starting Line

Maybe I Was Right, Maybe You Were Wrong
But True Facts = Equals Real Truth
Every Damn Time

If You Want To Dance
Go Right Ahead
If You Want To Play
Better Stay In Sync
Better Look In The Mirror
And Not Forget
To Be Sober And Use True Ink
If You Are Really My Friend
As Your Lies, Even When In Ink
Would Burn Under My Fire
Yeah Burn From My Fire

Questioning Your Judgements
Now? - Seeing Through
Your Self-Deception?
Because You're Shouting Me Gifts Again
That Would Never Quiet Me Down
I've Already Paid For My Sins
You Silly Clown
Don't You See?
The Pen Has Always Been
Always Been More Powerful Than The Gun
Can You See At All?

THAT PLACE:

Take Me To That Place
I Want To Go Once More
Don't Worry About The Face
I'm Not Keeping Score
Dancing With The Others
They're Trying To Stay In Touch
All Of Their Sweet Mothers
Have Had A Bit Too Much
Seeing So Many Places
And All Those Crazy Faces
I Sing And Laugh A Lot
Trying To Get Off This Spot
But Wind Up Repeating
All The While, I'm Meeting
Strange People And Great Customs
And A Number Of Severe Legends
Take Another Step
It's Time To Make A Move
But Why Should I Go
What Have I Got To Prove
Nothing To Nobody
Not Even To Myself
I'm Searching For Somebody
It Might Be An Elf
They Are Pretty Cool
Might Look A Little Green
Deception Leaves A Fool
Even Though They're Mean...

SPREADING THE FIRE:

(for Catherine)

I Was, I Will, I Still
Yes Still I Am
What You Would Say
Happy About Meeting You
On That Memorable Day

Make Yourself Comfortable
To Indulge And
Be Capable
Of Being The Happy
Beautiful Angel
You Are

Being Yourself
Not Needing Any Other
Simple Cliches
See Yourself For What
You Really Are
Yeah Not Needing
Any Silly Charades
Perfect Already
Yeah Already A Star, This Is True
So No More You Need
Yeah No Need To Be Afraid

When Your Fingers Start Crawling
Your Love Never Stops Spreading
Entered By Chance
Into My World
Then Into My Pants!
And You Still Put Me
Into A Trance

Your Subtle, Yet

Unforgettable Aroma
Just Enhancing Your Beauty
Always Enchanting My Mind

You Spread The Fire
From What I've Seen
Looking Into My Heart
Yes, - You're My Desire
And Always Have Been

Life, Love
My Friend
Here Tomorrow
Still Here Today
Some Good Things Never Go Away

You've Already Spread My Fire
From What I've Seen
I Think I May Have Always Known
You Are
My Desire And Always Have Been

Laughing With You
Always Chases Away The Blues
For It Is You
I Love Tomorrow And Today
As You've Seen It All
Yes Seen The Whole
True Game
Stood By My Side In Every Way

In Our Hearts And Mind
Is Our Love, Yours And Mine
With Us Always
Still Spreading The Flame
Until The End Of Time

TRUTH:

Change Is Consistent
Change Is A Fact
A Bit Unpredictable
Life Likes It, Like That
Try To Accept It, Try To Let Go
Slowly Or Quickly, I Try Not To Show
Becoming That Different Isn't Always That Bad
So Why Try To Help It
You'll Only Get Sad
A Natural Progression, A Thing Of The Past
A Perceptual Session
Only Coming In Last
Try To Avoid, Things Being The Same
You Know The Rules Of Life's Crazy Games
Learn To Compromise
You Don't Have A Choice
And Yearn To Hypnotise To One Day Rejoice
Love Me And Leave Me
I Love It That Way
Make Me See The Light, At The End Of The Day
No Two Days Are The Same

There's Nothing Identical
No Matter How Good
Whatever The Spectacle
Either Accept Or Regret
Times That Lay Ahead
Try Not To Fret
Just Sleep Where You Lay Your Head
Why Worry Over Nothing
It's A Natural Part Of Life
Only When It's Something
The Natural Is A Strife
Some Things Are Inevitable

They Are Bound To Happen
So Don't Get Uncontrollable
Just Learn To Slacken

    It's A Good Thing
    Boring Otherwise
    The Truth Like Spring
    There Is No Disguise
    Nothing Is Forever
    And Better Yet
    When Is Whenever
    In case You Forget...

'Knowledge Isn't Everything. Knowledge Is Limited.
Imagination Encircles The World'. Albert Einstein

ACT FIVE
WORK TO PLAY

WORK TO PLAY:

Take Life For Today
Don't Let It Go To Waste
Make The Memories Stay
Deliberate To Haste
I Hear The Morning Cry
The Birds, They Sing Then Fly
Catch Fish With Some Rods
And Overcome Impossible Odds
Let's Drink The Great, Sweet Wine
To Join Us When We Dine
Feel The Music Within
Life Too Can Be Devine

I Often Play The Fool
Jump When You Say Go
Race Within The Pool
How I Want To Take It Slow
Yes Fine Tune And Bloom
Fine Tune My Tools
What Ever You Desire
I'll Always Want More
Let's Sing And Then Regret
To The Places We Forget
The Times Of Our Great Friends
I Wonder If We Met
Believe In Resurrection

Keep Potential Alive
Because In The Sea Of Damnation
Only The Strong Minded Survive

When The Fun Is Over
How You Want To Sing
Life Then Can Then Surprise You
Surprise To Anything
Look Up To The Sky At Night
See The Shining Light
Life Again, I See
Infinite And Free
The Potential Vast And Silent
Hope We Aren't To Late
As We Work To Play
Yes Sometimes We Work To Play
Play Out Our Fate

Know Your Limitations
Life's The Only Thing I Own
I Know Not What I Want
But How I'm Nearly Grown
Take Me As I Want It
Don't Give Me All That Sort
Of Rubbish And Assorted Shit
Life Is Way Too Short…

WORK FAST/FAST WORK:

I Want To Wake Up
Wake Up In Your World
Today, With No Pain
And Tomorrow With Little
To No Heavy Rain
But When You Want It
Yeah It Goes Away Fast
And When You Hate It
It Always Seems To Last
Yeah Sometimes We've Gotta Be That Fast
Yeah Work Fast

Just Back To Work For Me
I've Already Finished My School
Yeah Already Finished Swimming
Yeah I've Already Fished That Pool
Looking For More Answers
Trying To Get On Back
Yeah Trying To Get On Track
Look A Little Closer
Yes You're Sure To Find A Crack

Go From Here
From Here To There
Don't Be So Weird
Yeah You'll Go Anywhere

Look At Their Faces
Simple, Yet Confused
...And All Of These Places
Yeah So Ironic, But Still Amused
Out The Back Door
Getting Sick Of This
Going Back For More
Yeah I Wouldn't Like To Miss

Tell Me The Answer This Time
What Am I To Do
Find Every Pleasure
Yeah Every Pleasure From Me To You
Now Let Me Go Home, My Work Is Finished
Yeah Just Go Home
Where Is Home For Me
The Jungle, City Or Highway
Or Just Running Free
Yeah Just Roaming Free

Take Me There, But Don't Lose Me
Yeah Take Me There, But Don't Please Me
Just As Long AS I'm Free
There's Nowhere Else I'd Rather Be
Waking In The Morning
The Best Time To Awake

Kind Of Like Spring
Yeah The Spring I Want To Take
Take It And Then Leave It
Yeah Leave It, Then To Sprinkle It
Grab A Chair, Then Decide To Sit
Anywhere Hey, This Is It

Are You For Real
Yeah So Hard To Forget
How You Made Me Feel
But No Bitter Regret, Cause
Nobody Knows Me
Knows The Ins And Outs
Yeah Just Let It Be
Yeah, There Or There Abouts
Confused Or Mesmerised
Don't Be Such A Sook
Lost Or Maybe Hypnotised
You've Gotta Know Where To Look
Yeah You've Gotta Know When
When To Read My Work Book
Yeah My Fast Work Is Where To Look
My Fast Work Is Just
My Own Hook.

THINK FAST:

Sometimes? No Always
The Truth Must Be Told
And If You're Still Reading This Book
You Have Been Chosen
Yeah Chosen As Instruments To Achieve
And Look...
Yes To Achieve The Real Objective
As You Are All, Already And Mighty Bold
And This Doesn't Have To Be Sold
Just Told...

The Truth However, Must Not Be Represented As
Truth
Too Many People Who Are Needed In The Struggle
Will Simply Die
...And All The Others Will Just Constantly Cry

The Cover Of FICTION Must Be Used
To Present This Truth
Those That Fear The Light

Will Not Want To Bring Attention To You
By Allowing Your Death
This Is The Only Way To Win This Fight
Yes The Only Way To See Real Light

This Is The Only Way
Do Not Be Afraid
The Fight For Real Humanity Demands Your
Courage
So Do Not Let Anything Or Anyone Discourage
Your Quest, As Tonight We All
Yes We All Will Need A Very Good Rest
As Our Future Is Happening In Our Past
And To Win This War
Yes To Win We Need To Think Very Fast
Yeah, Very Fast
Not Last..."

(SOME OF YOU WILL REMEMBER WHERE
THIS QUOTE COMES FROM AND OTHERS
WILL NOT! ✍)

NOT JUST A PAWN:

There She Goes
She's Left Us Now
Watch Her Nose
Then Watch Me How
I Unravel Her Mind
Break All The Shields
What Things Can I Find
Then See What Yields
I Possess The Power
To Break Her Little Heart
Right Within The Hour
Right Through From The Start
But That Is Not The Case
Well In This Lifetime
I'm Pointing At The Wine
She Can Play Her Game
For I Am Just A Pawn
She's The Power Of The Queen
Just Like A Newly Born

The Go Is
Hell, I Don't Know

It's Starting To Fiz
Now Watch Me Go
My Blood Starts To Boil
My Mind Goes Ahead
Like In Burning In Oil
Am I Better Off Dead
At Least For Now
Just Watch Me How
I Eat The Cow
Hey, Like Wow

Triumphs Big
Triumphs Small
Big Or Small
I'll Eat Them All
Not Just A Pawn
No, Not That Call
My Hunger's Gone
Now I Begin To Itch
Watch The Curse
Of The Nasty Bitch

SHOT DOWN:

You're Still Living Like
No One Ever Forgets
Or Living In The Common Ground Of Regret
Still Playing With Fire
Yeah Still A Liar
You're Shot Down
Even When Hiding In Your Lair
We're Coming After You
With Our Own Guns
Yeah With The Truth
And A Pen
The Strength Of A Bear
Will Tear You To Shreds

Shot Down Again
Yeah You're Shot Down Again
Should Of Never Left Your Kennel
Better Keep Running, Dog
Time Is Nearly Up

Yeah You're Shot Down Again

Don't Try Any Evil Games
War Against War
You'll Never Win
We've Always Been In Front
Of All Your Sins
No Where Now To Hide
Just Go And Decide
Which Side You're Really On
Or We'll Just Shoot You Down Again

Shot Down Again
Yeah You're Shot Down Again
Should Never Of Left Your Kennel
Better Keep Running, Dog
Time Is Nearly Up
Yeah You're Shot Down Again

SUPREMACY KNOWN:

He's The King Of The Sky
The One With No Bound
He Speaks With His Mind
But Then Makes No Sound
Supremacy Known
And Proudly Held
Humbly Returns
Unescapable Ground
Strolling Along
Scratching Firm Hoping To Come Upon
A Wriggling Worm
Persistence Rewards
He Takes To The Air

Carefully Now
He Answers Their Prayer
Nesting Among
Those Close To Breath
She Nurses The Young
And He Protects
Obliged Without Fail
The Cold Night Grows Near
His Wings Are Nearby
To Help With Their Fear…

NOWHERE:

You Know What They Say
You Cannot Buy Love With Money
Well Love Without Pay
Is Pretty Damn Funny
They Say Materialistic Child
Oh, Your Attitude Needs A Rest
With Money You Are Quite Wild
And Heaven's Simply The Best
What's Money To Us Anyway
Plain Old Paper
The Plastic, Numbers, Various Things
That Turn Us Into Anything
Oh No, We Have Our Morals
Scruples, Obligations For A Start
Oh No, The Little Florals

Hey, I Do Not Want To Part
So, I'm A Little Plastic
Well, I Never Said I Wasn't
Or Some Kind Of Spastic

Who's Suffering The Tick
Yeah, I'm A Little Crazy
Today, The Modern Age
Maybe A Little Bit Hazy
Tomorrow Comes The Rage
In France They Are At Peace
The Glorious City Of Love
Does Love Pay The Bills?
No, But It just Needs a Shove
What Are You Talking About
Oh Love
Well Hey, Don't Get Me Wrong
I Think I'll Sing This Song
Then Maybe I'll Forget
The Singing Of This Set

If REAL LOVE Comes Along
Grab It For An ETERNAL Song
Because If
Money Gets In The way
You'll Have Nowhere To Stay….

TOO SLOW:

Who am I
I am Who
Just a regular guy
Just like you
Without the Glamour
No tin foil
Give the hammer
Or the foil
For I'm not fussy
I'm not sad
I'm just easy
Then maybe glad
Give me a break
With no heartache
Better than that
I'll have a kit kat
Yes, fortunately

That is in stock
Normally
I'm a rock
Without the edge
No big thing
Try and wedge
Me to sing
Even a ring
Will give me more

More than I expected
Lick the sore
I couldn't be bothered
No mucking around
I go with the flow
In order to sound
Happy and sad
I'm there when you need me
I'll give you the seat
'cos when we're all alone
No dog or bone
We can say or sing
Most everything
Trust plays a role
In kicking the goal
In getting the ball
Look out for the pole

Hey, the jacks
They get here so fast
No time to blink
It's in the past
Forget or regret
Move over, let's go
I'm beginning to sweat
Oh no, Too Slow…

CHOICES:

Contemplating Life
And Retrospective Views
Help Me With The Past
To Help Find All The Clues
Looking From The Bottom Leaves Us Lying In The
Rain
The Only Way Is Up
Yeah How Can I Refrain
Stretch Me All The Way
Away From The Black Edge
So Close To The Bottom
Yeah Just Like A True False Pledge
Do I Have A Choice
Yeah I Decide To Play
We Always Had A Choice
And It Works Out Best This Way:

We Are The Ones Who Will Face It All
Yeah We're The Ones Already In Our Own Stasis
We're The Ones With Real Faces
Yeah We're The Ones
Yeah Already Strengthened Our Bases
We Are The Ones Using Our True Voices
We're The Ones With Our Own Choices
Yeah Shout Our Own Choices Now
Use Them Now, Loud
Yeah Not Then
Forever, Shout
Yeah Never Lose Them Again
Yeah Shout
We're The Ones Who Don't Fake It
Yeah Even If We Don't Make It
We're The Ones, Always Facing It
Now

We're The Ones Who Create It
Yeah Always Create It
Louder
So Now So Just Fly

Don't Plan To Make Believe
Leaving No Place To Hide
And Don't Plan To Deceive
To Reach Any Side
Choose Wisely This Time
Yeah You - Just Decide:

Oh, It's Good To see You
Yeah Haven't Seen You For So Long
Been Busy With Accomplishments
Yeah, Hope That I'm Not In The Wrong
You're Looking So Much Better
Such A Low Descent
You've Gone Through So Much Trauma
Yeah, What A Compliment

Each To Their Own
It Works Out Better This Way
Do We Have A Choice
Yeah So I'll Always Decide To Play:

Hey There, What You Doing?
For You Are Just Not Right
People Are Mysterious
Yeah They Don't Understand
The Places We've Gone To
The Customs, All Of Ours
Yeah The Customs Of Our Wild Land

So Superior Are You?
Out Of Your Mouth Comes The Law
So How Can I Therefore Contradict?
Yeah Contradict What I Saw
Yeah What We All Fucking Saw
We Are The Ones Using Our True Voices
We're The Ones Sticking To Our Own Choices
We're The Ones Who Don't Fake It
Yeah Even If We Don't Make It
We're The Ones, Always Facing It
Now
We're The Ones Who Create It
Louder
Yeah So Never Debate It
So Now Just Fly

One Thing I Have Learned
And This I Have Learned Slowly:
Pay No Heed To Others
As A Lot Think They're Holy
Take Care Of Yourself First
And Get Back On Your Feet
You Have To Take A Step Back
Off This Seat
Yeah Off This Perceptual Seat
Your Potential Is Enormous
Don't Ask The Question Why
Just Go Ahead And Do It
Now And Loud...
The Sky Is Yours
So Fly...

MOTIVATED:

Laughing With You Fool
Hey, I'm The Fool To Be
You Think That You Are Cool
Well, Laugh Again With Me
The Red Draws To A Close
They Want Me To Impose
The Wild And Savage Beast
They're Dreaming Of The East
Collect The Inner Thoughts
And Put Them To The Test
Work Out Which Goes First
There Really Is No Best
Motivate And Work
Your Art Collectively
Chase Away The Blues
Populate Respectively
In Another Time
I'm Choosing Different Wine

And The Ones That Are
The Simplest Here By Far
Choose The Same as Thee
Choose The Same As Me
Hey, You want To Be
Exactly
Like All The Rest
You Are The Best
And Unlike Some

In The West
My Dreams, Do They Work Out
I Want To Then Get Out
Then To Then Unfold
And Sing Here In The Cold
Like I'm Getting Old
Well, So Am I
In The Freezing Cold

The Dancing In The Sky
Is Just One Great Big Lie
But How Can You Tell
The Breaking Of The Spell
In The Inner Light
I Think I'm Turning White
Then Again I'm Not
Getting What I Got
Then Again Pretend
This Is Utter Rot…

GO BEYOND...

Set Me Alight...And Forever It Seems
Fear The Fright Of Your Wildest Dreams
Violence Is A Natural Act
Red To Black
Yeah Replace The Dreams
This Is A Fact: Yeah No Distract
Start To Go Beyond
Yeah Beyond... And In Even The Dark Pond
When You See All The Light, You'll Ask
Is This Right?
Yeah Is This My Fight?

Fact And Fantasy United As One
Real Power Stems From The Barrel Of A Pen
Living Memory Can Easily Be Erased
Easy To Be Replaced
Yeah The Pen Is Mightier Than The Sword
Yet With No Blade, Just A Lot More Power
Yeah More Power Than The Gun

You Wonder What's The Scam
Pool, The Water, The Sand
And In The End There Is No Land
Only Freedom Here And There, But
Sometimes Hate Is Their Law: Love Under Will
And...
Transfusing, Contagion Breeds
With Slight Of Their Firm hand
Is All Deviancy Cleansed? Or Just Another Order

For The Next Kill

Yeah With A Deceived Sleight Of Hand
A Diseased Slight Of Hand
A Multi-Cultured, Sometimes Evil
Epidemic Can Be Released

When You Think Of It Some More
Your Head Starts Getting Sore
And Wind Up With Regret
Yeah On Your Shirt Lies The Blood And Sweat
Bitterness
And The Lies – Don't Believe Them All
Lies -Yeah Don't Swallow Them Whole
Yeah Lies, Half Truths, Misinformation And...

It Comes And Goes...And Soon Creeps Up On Thee
After All The Weal And Woes
We Just Want To Be Free
Yeah Chase Them Far Away...Away And Let Them
Stay
Into The Outer Day, Yeah
Just In Their Own Ashtray
Hell, I'll Always Stay And Tomorrow
Is Another Day; I Want To Go Away
But Help Me Here To Stay
Never Pushed, Never Scared Today
Yeah I'll Never Go Away

SNAKES & LADDERS:

Sometimes We See Them
But Still Manage To Trip
Trip Over, Some Of The Left Over
Things That Belong In A Grave
Or In A Rubbish Tip Full
With Nothing That Can Be Saved

Sometimes We Hear Them
And Still Don't Listen
To That Special Part Of The Brain
Watch Out, They're Everywhere
Mixed Well And Within All Our Realms
Yeah Even In The Rain
But I Just Feed Them Left Overs
From My Own Bin
Sometimes Even Scraps, Depending On Levels
Of Stupidity, Ignorance And Skin
Yeah Feed Them
From Their Own Fucking Rubbish Bin

Aim High, Never Low
No Middle Ground, Sometimes Even Tip Rats Try
And Hide Even In The Snow
Yeah Sometimes Snakes Somehow Enter And Hide
On Your Favourite Movies Or TV Show
Real Ladders, Never Break
Never Shatter
Yeah Real Ladders Give
Give You A Stable
A Different Height
The Real Edge Of Flight

Yeah Let Yourself Listen To The Real
Music
And All The Different Strings
Yeah To The Right Sound
As Long It's The Truth You Seek
No Ladders Will Ever Throw You
No Snakes Will Ever Drop You
Yeah Never Will They Matter
Just Choose The Good Food
From Your Own Platter

Now Their Venom Is Just Poisoning
Themselves, They're Lost Forever
Yeah Never Found
Cause Now You're Not Blind
And Your Mind Is Alive
Yeah You've Found, Your Real Ground
Whether In Front Or Behind
They Will Always Just Fall And Shatter
Don't Even Have To
Think About The Latter
Never To Surface Again
No Tears To Be Shed
Cause Now You're Eyes Are Actually Open
My True Friends
Yeah, Now Really Using Your Head
Just Like I Always Said
It Wouldn't Take Much For You
To Aim High, Stay High And Always
Keep Ahead

ACT SIX
Sleep & seed(s)

SLEEP:

I'm Absolutely Tired
So Tired, I Can't Think Straight
That Wired, Waiting Is A Breeze
So I'll Retire To A Freeze
Catch The Slim Chance
Of Looking At A Glance
And Then DO Nothing
With The Chance, I Am Hoping
Sleep

The Only Thing Right Now
The Only Thing To Me
Rest And Watch Me How
To Sleep Endlessly
You Know That Time Of Day
When All You Want To Say
Is Goodnight And Goodbye
I'm Off To Bed To Die...

PLEASANT DREAMING:

Yes, Yes And Yes
One More Time
Consume And Feed Me
With The Wine, Come Again
Or Will You Go
Come And Then
Hell, I Don't Know

If You Pay, I Will Not Stay
But If You Go, I'll Take It Slow
Well, At The Time Anyway, At Least
Save The Wine For The Feast
One, Two, Four - You're Out
But I'm Still In, Where's The Shout
Hunt It Down
Hunt It Cold
Even With The Dog
You Know I'm Sold

Again, Again
Oh My, Yes
When And When
I Missed My Guess
Well, How Were You To Know
I Couldn't Help That
Remember I Am Male, Where Do I Fit
Upon The Scale
In The High
In The Low, I'm Sure To Be
In The Red
Come And See

I'm In All White
Hey, So Are They
And In The Night, You Have To Pay
The Screams And Carry On
I Said Let's Just Have A Bong
With This They Were Mystified
And Puzzled With This Song
Pleasant Dreams....

DREAM IF...

Right Up
Right, Up There
They're Dancing In The Air
Help, A Freak
It Wants To Speak
Well, Send It There
Healthy Like Before
We'll Try And Run All Day Yeah, Now I'm Wanting
More
And Maybe Want To Stay

Jump
And Look Ahead
Thump
I'll Walk Instead
Looking All Around, I See Them Now
Round And Round
Hey, Like Wow
Beaten, Have No Edge
But Follow Up So More
Here, Here's A Pledge
To Settle Up The Score
Thank You Sir, You Are An Ass
With Your Cool Fur
Asleep In The Grass

Sleep On The Grass
I Will And Will Indeed
Watch Out For The Snow
I'll Even Plant A Seed
And Hope That It Will Grow
Keep It Steady, Be Prepared
Are You Ready
I Am Shit Scared

Yeah, Feel It Go
Hey, I Want To Know
Are You There, Say Hello
Come On Now, I Want To Know
Selfish Aren't We, Don't Do It Again
Drifting Free, But Until When
How Long Will I Stay Here
Or When Will I Go
Is The Floating Deer Saying
Let's Go
Cows And Deer
What's The Diff
Both Have Tears
But Do Both Think If...
Do Both Dream If...
Yeah, Just Dream IF..

BEYOND THE HORIZON:

A Raindrop Here
A Flower There
A Sense Of Magic In The Air
Cool, Calm Breezes
Passing By
Trying To Catch The Butterfly

Swift As Arrows
The Eagles Fly
A Lizard Hanging In His Joyous Cry
Beyond The Horizon
The Sun Shines Bright
Spreading Colours To Our Bewildered Sight

Over The Hills
Where The Deer Roams Free
A Miracle Is Soon To Be
A Mother Breathes Life
In To A Young Newly Born
Striving For The Perfection
Of A New Day's Dawn..

TRUST:

Trust Isn't Easy
Trust Isn't True
A Little Bit Sleazy
To Both Me And You
For We Are Just Flowers
We Are Just Human
Knowing Some Hurts Me
Time And Again
Low And Behold
Tight And Secure
We Grasp At The Old
Escape From The Lure
But Then To Entice
It Looks Very Nice
Should I, I Must
Huh, A Matter Of Trust...

Oh Well, What The Fuck
I Don't Know
Which The Way
Or Where To Go
Fast Or Slow
Far Or Near
You'll Never Know
If You Never Go...

SEEDS:

Some See A Realm Of Chaos And Night
Yeah Some Seed A Realm Of Just Fright
But We're Here To Tell You
That Death Isn't Anywhere Tonight
That Devil Has Left The Stage
Yeah Left His Own Stage In Chaos And Fright

Yesterday Was A Million Years Ago
In All My Past Lives I Must Have Been An Asshole
Yeah Just Been The Arsehole
And All I Did Was Spill Some Good Seeds
At Your Children's Feet
This I Conceit, No - Never Discreet
Just Helped Spread
Yeah I Spilled Some Right Seeds At All Of Our Feet

We'll Never Say Goodbye
Even When We Lie Down
Yeah Even When We Die
As Now Is Not The Time
Yeah Now Is Not The Time
To Even Cry

Seeds Of Thought, No Distort
Seeds Of Logic, Yeah No Problem
Seeds Of Truth - Yes Just Be Bold
Yeah Just Be Bold Even In The Cold
Seeds Of True History And Science
Yeah No Defiance
Just Water Their Seeds
Yeah Water Their Reliance
Yeah Just Reward Their Compliance
With Justice: A Dish Served Best, Hot Or Cold
Yeah Justice; All Of It Will Hold

Chaos And Disorder Is What We Don't Have
Yeah Not Today, That's What We Should All Ask
Yeah Just For One Day
Love, Truth And All Else Needed
Is What We're After
But, Yeah Just In So Many
So Many Different Ways

Yeah Come And Take Me Baby
We'll Try And Steal The Night
Capture It And Hold It
Yeah And Take It With Our Flight
We'll All Travel Down The Road
See All There Is To See
But Take It As It Comes
Yeah We Want It Naturally

Let's Pick All The Good Fruit
It's Ours And Ours To Reap
Yeah Let's Guide It, Do No Wrong
And Hope All The Hierophant's Are
Yeah Are Not All Asleep

The Hermit's Are Still All Here
Yeah Not Everyone's Still A Fool
Even If You Are, It's Still, Definitely Cool
...And The High Priestess Will Still Cradle You
Yeah Rescue You In Any Damn Drowning Pool
Yes Rescue You From Any
Of The Cruel

Yeah Remember
We'll Never Say Goodbye
Even When We All

Are Still Lying Down
Yeah Even When We Die
As Today Is Still Not The Time
Yeah Now Is Still Never
Never The Time To Just Cry

Seeds Of Thought
Still No Distort
Seeds Of Logic, Yeah Still No Problems
Seeds Of Truth, Yeah You Have Always Been Bold
Seeds Of True History And Science
Yeah Still Now, No Defiance
Just Keep Watering
Yeah Watering Their Seeds
Yes Watering Their Lives
And Never Forget
Yeah Never Forget Our Children's
Reliance...

LIFE SHADE:

I'm Developing A New Thing
My Present State Of Mind
Is Constantly Searching
Searching For Something New
Yeah Something True To Find
A Change Is As Good As A Holiday
Well, I'm Going Through A Phase
One Day I'll Be Back
Life, Like Entering One Big Maze

I'm Entering A Different
Quite Interesting Experience
I Don't KNow What It Is
But It's Making Quite A Difference
To Both My Unregretful...And...
Regretful Life

The Life I Choose To Live
My Shape Or Your Viewpoint
Make Me Want To Give A Certain Shade Of Light
Allows You To See Within...One That Seems Alright
But You Can Never Win
I Bet You Didn't Know That
I Illuminate My Life
So You Can Only Find The Scratches On The Surface
Are A Deeper, Stanger Kind
Reality Can Be So Unreal

Did They Ask You That
But Please, Won't You Tell Me
How Did It Make You Feel
In A Given Point In Time
I Might Be Something Else
Not Somewhere Or With Someone
And Not Even A Mouse

It's Sometimes Clear And Oblique
Yes That's A State Of Mind
Still, Unique
If It's Something New I Find
I'm In A Later Stage
Change Is Still My Friend
No Matter What The Question
With Me 'Till The End

My Colour Hasn't Changed That Much
My Face Still Looks The Same
So Why Is My Mind Playing A Different Game
Rules And Regulations
They're All The Same To Me
They're Mind Manipulations
To One Day, Set You Free
Find The Answer To The Question
The One That Fits Right In
Make It Act Like The Solution

Then The Answer To The Sin
Yeah Then The Answer - To The Bin

Come Drink And Dine With Me
Intoxicate Within
We'll Have A Feast Or Two
Yeah, And See If We Can Spin
Understand Adventure
Explore To The Unknown
Pioneer Your Own Life
Freedom Is Yours
Yeah Freedom Is Yours To Own

Sometimes I Get The Urge
Yeah That Urge, The Feeling Is So Great
Yeah Escaping From The World
To Find Another Fate
That Is What We All Want
Yeah No Problems Or Regrets
To Keep Our Spirits Free
Oh My Dear Angel, IF Only..

WAKE UP:

One More, What The Hell
I Think, Sneeze And Cough
I Think I'm Not Too Well
The Swelling Doesn't Seem Unlike The Perfect
Dream
We've Seen It All Before
The Dream Is Getting Sore
I'm Getting Sick Of All This
Today, I Wonder What I'll Miss
Hey, I Don't Really care
Hey, Look Out, It's The Bear
No, I Can't Go On
With All This Carry On

I Turn And Twist And Shake
My Mind Then Doesn't Like The Twisting Of The
Ache
The Turning Of The Mic
I'm Shaking All Around
The Pictures Of The Sound
Where's The Common Ground
What The Hell, Wake Up..

ACT SEVEN

FOREVER FLOWING:

SHE STILL TAKES MY BREATH AWAY
EVEN TODAY
AND EVERY OTHER DAY
EVERY ONE OF
ALL OF MY THOUGHTS
STILL FLOWING STRAIGHT
MOVING EVENLY
QUICKLY IN THE RIGHT DIRECTION
YES IN THE REAL DIRECTION
INSIDE
INTO
MY HEART
STILL FEEDING
STILL FEELING IT STEADY
FLOWING LIKE A RIVER
RIGHT FROM OUR START
YES NEVER TO DEPART
FOREVER FLOWING, FOREVER KNOWING
THAT
NOTHING WILL EVER TEAR US APART
YES FOREVER KNOWING
RIGHT FROM MY HEART

I STILL NEED HER
I STILL FEEL HER EVERY DAY
WHISPERS OF LOVE
KEEP ME INFECTED
HER WHISPERS OF LOVE
KEEP ME CONNECTED
EVEN SILENT WHISPERS FROM
HER HEART WILL BE COLLECTED

A CONNECTION LIKE NONE BEFORE
SOMETIMES STILL FEELS LIKE A DREAM

SOMETIMES STILL SEEMS LIKE THAT
GOLDEN AGE DREAM
BUT IF IM REALLY SLEEPING
NEVER WANT TO WAKE UP
I KNOW IM NOT DREAMING
AS DREAMS ARE NOT THIS PERFECT
YES REAL LIFE IS FUNNY SOMETIMES
AND THIS IS AS REAL AS IT GETS
CAUSE WHEN YOUR LOVE IS FLOWING
FLOWING THROUGH MY SOUL
I KNOW IM AWAKE
AND FULLY IN CONTROL

NO LONGER CAN I BE DECIEVED
BY ANY ENEMY'S AROUND ME
CAN'T BE CONTROLLED BY ANY
ANY IN CLOSE PROXIMITY

AS I ONLY NEED YOU
AS I HAVE YOU
YES, I HAVE THE TRUTH ON MY SIDE
AND AS LONG I KEEP YOU ? TWO
MY SOUL WILL NEVER DIE
FOREVER FLOWING, MY LOVE
FOREVER YOU, MY STEADY ROCK
ONLY YOU BEAUTIFUL
AND THATS NO CROCK

YOUR TRUE NATURE
MORE CALMING THAN AN ISLAND
BREEZE
A CALMING BREEZE RIGHT INTO MY FACE
ALWAYS SOOTHING
AS YOUR HEART IS LARGE

WHEN IM AROUND YOU
THERE IS NO RAGE
JUST A SMILE ALL OVER MY FACE

NOW I KNOW
BUT ALWAYS BELIEVED
YOU ARE MY ANGEL
AND ALWAYS WILL BE

NOW WE ALL KNOW
THEN NOW, YES NOW WE ARE FREE
AND THE FUTURE WILL BE FLOWING
NO NEED TO STRESS
ABOUT NOT ENOUGH TIME
OUR FUTURE WILL BE FLOWING
UNTIL THE END OF TIME
SOME MAY DISAGREE
BUT IN THE END, ALL WILL JUST
PLAINLY SEE
THAT SOMETIMES THESE THINGS WERE
MEANT TO BE!

SOULMATE

NOW WE KNOW WE TRULY EXIST
NO MORE HUGE CHALLENGE
WILL TRULY BOTHER
AND PERSIST.

TRUE LOVE BOUND
BY STRINGS
YEAH NOW UNBREAKABLE COSMIC
STRINGS

SHE'S MY BEAUTIFUL ONE
AND ALWAYS HAS BEEN
NOW ITS HER TURN
TO BE TRULY HEARD AND SEEN
DON'T HOLD BACK MY ANGEL ?
NEVER HOLD BACK AND THEN
YOU WILL TRULY SING
JUST SIT BACK AND WATCH WHAT IT
BRINGS
MY ONLY

MY ONLY UNIVERSAL QUEEN
YEAH MY ONLY TRUE SOUL MATE

NOW THAT SHE'S TRULY SEEN
YEAH NOW THAT'S SHE'S TRULY HEARD
SHE'LL NEVER BE SHUT DOWN AGAIN
YEAH NEVER HAVE HER WINGS CLIPPED
AGAIN
LOVE IS INFINITY, YES THAT'S TRUE
SO FLY AGAIN NOW MY ANGEL ?
WE ARE LIVING PROOF
ME AND YOU ?
TWO ? HEARTS AND ONE MIND
BEATING AS ONE
I LOVE YOU MY CATHERINE

ALWAYS AND FOREVER
AS THERE'S NO SUCH THING AS
COINCIDENCE, : - JUST OUR DESIGN
YEAH JUST OUR DESTINY AND FUTURE ?
ALL TIGHTLY WRAPPED UP
IN CONNECTING STRINGS

YEAH THERE'S NO SUCH THING AS
COINCIDENCE,
ITS JUST MEANT TO BE
ITS OUR CONNECTION
KICKING RIGHT IN
RIGHT THROUGH THE DOORS OF SPACE
TIME.
YEAH RIGHT THROUGH THE THIRD,
FOURTH UNTIL THE LAST
OF ETERNAL LIFE, LOVE, KNOWLEDGE AND
TRUTH.

YEAH
CONNECTED FOREVER
ITS ALL JUST SYNCHRONOUS MATHEMATICS
THROUGHOUT THE WHOLE
OF THE CREATOR'S
UNIVERSE

JUST SMILE ? NOW MY BEAUTIFUL ONE
AS WE HAVE ALREADY WON OUR LAST
BATTLES

AS NOW WE HAVE PROVEN THAT LOVE CAN
BE MEASURED
AS A SEPARATE ENTITY

YOU WILL ALWAYS KEEP AHEAD

SMILE:

My Heart Beats So Fast
When You Love Me Like You Do
Breathing In, Breathing Out
Never Skipping A Beat
Always In Tune
Now That's Real Timing....
And Real Poetry
When Touching Me
Feeling Me Like You Do

Sometimes Our Heads Need Shaking
Anywhere, Anytime
Never Ending
Un-explainable Connections Of Yours To Mine
It's Just Our Time To Shine
Yeah, Really Shine

Naked Skin On Skin
With You I will Never Feel The Cold
Keep Feeding Each other
Our Heart Connection
Yeah, Our Heart Correction
Just the same Old Soul Mate Corrections

Every time our eyes meet, You and I
Feel That Sexual And Emotional Trance
It's No Bullshit Stance
Its Just a Real Love Trance
Every time our heart beats
When We Close our Eyes
When We Open Our Eyes
It Just Takes Our Breath Away

I Love Your Smile

Never Want to see it go away
I Love Your Laughter
Caressing My Soul
Just Take A Chance On Me
I Just Want To Feel Your Tears Of Laughter
That's A Promise I Will Keep

One Day We'll Have Our Own Story
But A Billion Times Better Than Any Movie
For You Are One In A Billion....
And My Own Personal
Shooting Star

I Love How You Tease me
I Love How You Taunt me
I know you're Hoping and Praying
But All You Need is
Faith
In Love

I will Never let you Slide or Fall
From Your Stable of Where You Are Right Now
All I'm Promising Is More
Cause As My True Soul Mate
I Owe This to You

That Vibrating Connection Is Real
A Real Reason Why It's There
I Know You Know Me
I Know You Love Me
And I Know That In Your Heart
You Trust Me
Just A Leap Of Faith Is Needed
Like I Already Said

You Won't be Falling Anywhere
Unbreakable Ladders Of Love And Hope

Helping
Other
People
Evolve

Just Evolve With Me
And Then You'll Be
Truly Free

As Your Eyes Are Real And True
Not Just Blue
Real Eyes See Real Lies
You'll Always Have My Shoulder To Bare All The
Heavy Weights

Did I Forget To Tell You, How Much I Love It
When You
Tease Me?

SMILE! :)

EXPLOSIVE:

Just Like The Future Can't Be Predicted
Yes, Just Like Nature
Our Love Can't Be Restricted
Yeah, Our Love Can Never Be Limited

Sky's The Limit When I'm With her
Now I just need one breath of fresh air
Cause I just wanna stare into her eyes
Into her Soul
While Traveling on this Roller coaster Of Life
So YES
Just One Pause In Time, Get A Breath Of Fresh Air
Real Life IS
Yeah Real Love is just so EXPLOSIVE sometimes.

Our Connection never severs
It doesn't matter what room were in
Doesn't matter what dimension We're in 3,4,5 or 11
We are Living and Learning For The Next
And What Time Is It Now Actually?
Explosive Unknown Sensuality
Effecting All Parts and every Part
of our Bodies
Breathing Together
Touching Together
Skin on Skin
Love on Love = Explosive Even
In all 3 and 4 dimensions,
Yeah Which Only We can hear and see

She's So Explosive
Where did we meet in the Stars?
AEONS AGO

She's So Explosive Just seeing her that day
walking in the streets
She's so explosive, even when hidden under the sheets
So Explosive, Anytime, Anywhere
It doesn't Really Matter
Cause We Both have the world at our feet

I will never forget her beautiful heart
Never forget Her bluc eyes
Glimmering in the sun
As I always knew that day would bc fun
Yeah, just like shooting a big gun
But Bigger Explosive Fun

No more need to Hurry
No more need to run
Just walk with her
Talk with her
To feel her warmth through
Her own Hearts Eternal Sun
Yeah her Hearts Infinite Sun

Be Patient My Love
My Soulmate, and take my hand
Believe in me, that I'll always keep you safe and awake
Even if the world is at Stake
Take my hand
I'll Try To show you the real truth behind everything
You wanted to know
Soulmates Connected Again For Eternity
TRUE LOVE is EXPLOSIVE, it's a real String
Collision.
Yeah a real String Collizion.

TWO HEARTS//
ONE MIND:

SHE SAW ME COMING
YEAH, I SAW HER GOING
BUT AS ALREADY GOOD FRIENDS
NOTHING
YEAH NOBODY COULD STOP US FROM
MEETING AGAIN
IT DIDN'T MATTER HOW LONG IT TOOK
IN THE HEAT OR THE COLD
THE DARK AND THE RAIN
I KNEW SHE WOULD RETURN TO MY
ARMS AGAIN
INTO MY HEART AGAIN
YEAH SHE RETURNED TO MY EYES
THAT'S WHEN I KNEW OUR HEARTS WERE
BEATING AS ONE
YEAH THAT'S THE DAY I KNEW SHE WAS
MY BEAUTIFUL ONE

IT'S SO RARE WHEN SOUL MATES ARE
FLYING
BUT SOMEHOW SHE LANDED ON MY
SHOULDER AND WHISPERED TO ME
I JUST THANK THE REAL CREATOR
THE INFINITE UNIVERSE THAT MY EARS
WERE LISTENING
WERE REALLY OPEN THAT DAY

HER EYES AND HEART WERE ALREADY
OPEN
WE ARE JUST 2 OLD SOULS IN NEW BODIES
BUT WITH A FUTURISTIC STATE OF MIND.

1 BECOMES 3 AND 3 BECOMES 1
TWO HEARTS/ONE MIND
OUR LOVE
WILL NEVER BE BROKEN
A LOVE UNSEEN
AN INFINITE Love, Truth And Knowledge
A HOLY TRINITY OF THREE

EASY AS 123
TRUE HUMAN SPIRIT NEVER DIES
I MADE A PROMISE AND KEPT IT
PROMISES
NEVER BREAKING, NEVER CRACKING
SHE'S ON HER HIGHEST UNBREAKABLE
STEPS OF THEM ALL
THE HIGHEST OF ALL
2 HEARTS/ONE MIND
NO SNAKES TO BE SEEN AT ALL

THE HIGHEST QUEEN Standing TALL
SUITS ME WHO LOVES HER FOR ALL TIME
AND SPACE
AND STANDING BY MY SIDE FOR EVER
MORE
NO MORE SEARCHING
SEEKING
TWO HEARTS/ONE BEAUTIFUL MIND TO
GROW AND EXPLORE
SHE'LL ALWAYS KEEP ME FREE
AND I'LL ALWAYS GIVE HER REAL SITE
2 MINDS JOINING TOGETHER AT THE

RIGHT TIME
YEAH, SHE'LL SEE THE REAL LIGHT
THE TRUTH BEHIND IT ALL
TRUTH BEHIND MY ALL

THE AGE OF AQUARIUS IS UPON US
WE'LL BE TOGETHER TO SEE IT ALL
The BAD First, Then The GOOD
A New AGE To Unfold
ITS ALL LOVE
RESPONSIBILITY
FOR OUR CHILDREN'S FUTURE
THE CHILDREN OF PLANET EARTH
AND OUR UNIVERSE
HER MOON WILL SOON BE AGAIN
AND COMEBACK ALIVE
IN COLOUR
TO LIVE AGAIN

TWO HEARTS/ONE MIND
NO MORE HIDING
NO MORE FEAR
NOW OUR PATH IS CLEAR
AS SHE IS VERY CLOSE NOW NOT GOING
TO HIDE ANYMORE
LIVING JUST AROUND THE CORNER,
WON'T BE LONG NOW
WE CAN OFFICIALLY START OUR NEW
ROCK SONG
OUR NEW LOVE SONG
NO MORE HIDING THE BEAUTIFUL ONE
OUR NEW ROCK LOVE SONG.

ENO-LUF-ITU-AE-BEHT
ENO (ONE) LUF (LOVE) ITU (INTO) A (A)
EBE(H) T(BEAT) OR(HEART BEAT) =
ONE LOVE INTO A (HEART) BEAT

SOMETIMES ITS BETTER TO MAKE
YOUR ENEMIES BELIEVE THAT
YOU'RE AT YOUR WEAKEST, BUT
IN REALITY YOU ARE AT YOUR
STRONGEST.

TRANSLATION FROM "THE ART OF
WAR' BY SUN TZU

LOVE STRINGS:

ALWAYS BEEN THERE
JUST HARD TO FIND
AND HOLD ONTO

ALWAYS LOVE THERE
FLYING ? AROUND
AND IT LANDED IN OUR LAPS

SOME SAY IT'S A MIRACLE
SOME SAY ITS RARE
AND THIS MAYBE TRUE
BUT I BELIEVE WE'VE ALWAYS BEEN
CONNECTED -ME AND YOU

YEAH LOVE STRINGS SURROUND US ALL
JUST GOTTA DODGE AND WEAVE AND BE
AWARE
OF THE DARK STRINGS AND COSMIC
SNAKES
AS LIFE UNWINDS
SUCH IS LIFE -
ITS A HUGE CONFUSING MAZE
BUT IN THE END TRUTH AND LOVE

CONQUER ALL
YEAH THE KING AND QUEEN OF HEARTS
CONQUER ALL

LOVE STRINGS ATTACHED AND WEAVING
AROUND OUR LADDERS
LOVE STRINGS KILLING OFF HATE, EVIL ?
AND DEATH
LOVE STRINGS FOREVER UNBREAKABLE
AND CUTTING THE REAL HEAD OFF THE
SNAKE ?

ALWAYS MAKE SURE WHEN YOU FIND
YOUR LOVE STRINGS
TIE THEM TOGETHER IN A GOOD KNOT
So NOTHING Negative can get through
Yeah nothing can break you again
Break your Ladders again
Yeah UNBREAKABLE LOVE STRINGS
MERGES INTO YOUR LADDERS
THEN NOBODY WILL FALL APART
YEAH NOBODY WILL FALL AGAIN

Printed in the United States
By Bookmasters